01-WSM-2b

SENSOR TECHNOLOGY FOR WATER QUALITY MONITORING: FIBER-OPTIC BIOSENSOR

by:

Andrew C. Cannons, Ph.D.
University of South Florida

Valerie J. Harwood, Ph.D.
University of South Florida

2004

WERF

The Water Environment Research Foundation (WERF), a not-for-profit organization, funds and manages water quality research for its subscribers through a diverse public-private partnership between municipal utilities, corporations, academia, industry, and the federal government. WERF subscribers include municipal and regional water and wastewater utilities, industrial corporations, environmental engineering firms, and others that share a commitment to cost-effective water quality solutions. WERF is dedicated to advancing science and technology addressing water quality issues as they impact water resources, the atmosphere, the lands, and quality of life.

For more information, contact:
Water Environment Research Foundation
635 Slaters Lane, Suite 300
Alexandria, VA 22314
Tel: (703) 684-2470
Fax: (703) 299-0742
www.werf.org
werf@werf.org

This report was co-published by the following organization. For nonsubscriber sales information, contact:

IWA Publishing
Alliance House, 12 Caxton Street
London SW1H 0QS, United Kingdom
Tel: +44 (0) 20 7654 5500
Fax: +44 (0) 20 7654 5555
www.iwapublishing.com
publications@iwap.co.uk

Library of Congress Catalog Card Number: 2003107087

WERF ISBN: 1-893664-95-3 IWAP ISBN: 1-84339-677-7

University of South Florida

The research on which this report is based was funded, in part, by the United States Environmental Protection Agency through Cooperative Agreement No. CR-825237-01 with the Water Environment Research Foundation (WERF). Unless an EPA logo appears on the cover, this report is a publication of WERF, not EPA. Funds awarded under the Cooperative Agreement cited above were not used for editorial services, reproduction, printing, or distribution.

ACKNOWLEDGMENTS

The authors would like to thank the members of the Project Subcommittee for all of their feedback, comments, and encouragement. The authors appreciate and thank the Water Environment Research Foundation for the grant support that allowed this project to proceed.

Report Preparation

Principle Investigator:
Andrew C. Cannons, Ph.D.
University of South Florida

Co-Principle Investigator:
Valerie J. Harwood, Ph.D.
University of South Florida

Project Team:
Miriam Brownwell
Barbara Dodge
Robert Ulrich
University of South Florida

Project Subcommittee

Robin Autenrieth, Ph.D., Chair
Department of Civil Engineering, Texas A&M University

James Dartez
Royce Technologies

William J. Davies
U.S. Geological Survey

G. Shay Fout, Ph.D.
U.S. Environmental Protection Agency

Bennett K. Horenstein
East Bay Municipal Utility District

Darrell Jan, Ph.D.
Center for Space Microelectronics Technology, Jet Propulsion Laboratory

Jim Lazorchak
U.S. Environmental Protection Agency

Arthur Meyers, Ph.D.
Eastman Chemical Company

Andrew Ogram, Ph.D.
University of Florida

Water Environment Research Foundation Staff

Director of Research:
Daniel M. Woltering, Ph.D.

Project Manager:
Dean E. Carpenter, D.Env.
Eugenia Bruno Hart (from 08/2003)

ABSTRACT

The project team investigated two methods for detecting the pathogens *Cryptosporidium parvum* and *Helicobacter pylori*: a fiber-optic biosensor and a real-time polymerase chain reaction (PCR). Both employ recognition of specific deoxyribonucleic acid (DNA) sequences in the target organisms.

The Analyte 2000™ biosensor, originally developed to detect chemicals, uses a fiber-optic waveguide. This propagates a specific wavelength using an evanescent light wave (a surface wave that travels along the interface between the waveguide and the bulk solution). The light excites fluorescent molecules bound to the waveguide, but not in the bulk solution, theoretically enhancing signal while reducing background interference. Attempts to develop this system for the detection of DNA had failed because the target molecules could not be detected.

The project team designed a "sandwich-like" immunoassay for the biosensor. Specific oligonucleotide probes were designed to bind to the waveguides via biotin-streptavidin interaction and capture the target DNA, which were synthesized by PCR. The team then attempted to detect the DNA using a probe designed to bind to the target via two detection systems: (1) an indirect signal amplification system based on biotin-tyramide deposition; (2) and direct detection of the fluorescent signal from cyanine-5 (Cy-5) molecules. Little difference was noted between the signal generated with or without the target molecules. Signal was only of any significance when target amplicons were internally labeled with Cy-5 by PCR.

The team also investigated real-time PCR as a pathogen-detection method. Filtration of up to 1 L of well water followed by concentration and "cleaning" *H. pylori* cells by immunomagnetic separation, was used to detect *H. pylori* seeded in a water source. Following cell lysis, the extracted DNA could be used directly in conventional PCR targeting the 16S ribosomal ribonucleic acid gene to detect <265 cells/L of water. This detection limit is in line with the determined minimum infectious dose of between 10^2 and 10^4 colony forming units. DNA purification was not required for this level of detection. Initial studies to amplify lysed cells by real time PCR indicated that an incorrect product was made. When purified DNA was used for real-time PCR, the correct product was produced from DNA representing as few as 100 cells per sample.

BENEFITS

- Evaluates the use of a biosensor, the Analyte 2000™ biosensor for rapid detection of pathogens in water.
- Describes methods that use DNA:DNA hybridization for the detection assay.
- Describes the exhaustive experiments that were performed to establish that the Analyte 2000™ biosensor was not effective for use at the DNA detection level.
- Describes a method for the rapid and sensitive detection of *H. pylori* in well water using PCR and real time PCR.

Keywords: DNA:DNA interactions, biosensor, PCR, *Cryptosporidium parvum*, *Helicobacter pylori*

TABLE OF CONTENTS

LIST OF TABLES

LIST OF FIGURES

LIST OF ACRONYMS

BHI	beef heart infusion broth
BLAST	Basic Local Alignment Search Tool
BSA	bovine serum albumin
BTS	biotin tyramide signal amplification system
CFU	colony forming units
cpm	counts per minute, unit measuring radiation
Cy-5	Cyanine-5
DNA	deoxyribonucleic acid
ELISA	enzyme-linked immunosorbent assay
FBS	fetal bovine serum
FITC	fluorescent isothio-cyanate
GC	guanine and cytosine
GlmM	gene encoding phosphoglucosamine mutase (old Urease C) in *H. pylori*
HRP	horseradish peroxidase
hsp70	gene encoding heat shock 70 protein in *C. parvum*
IDT	Integrated DNA Technologies
IgGi	immunoglobin G helibaceter cryptosporidium
IMS	immunomagnetic separation
LOD	lower level of detection
PBS	phosphate buffered saline
PBST	phosphate buffered saline, 0.1% Tween 20
PCR	polymerase chain reaction
QCM	quartz crystal microbalance
RAPD	Random Amplified Polymorphic DNA
rRNA	ribosomal ribonucleic acid
SSC	standard sodium citrate buffer, pH 7.0

EXECUTIVE SUMMARY

Rapid detection of waterborne pathogens, such as the bacterium *Helicobacter pylori* and the protozoan *Cryptosporidium parvum,* is essential for the timely and rapid identification and treatment of contaminated water to reduce health risks and associated costs. Current methods use techniques that are typically time-consuming, require monitors to have extensive training in microbiology, and can give delayed results. This project investigated the use of a fiber-optic biosensor assay to rapidly detect microbial pathogens that are of concern to the water/wastewater industries. The Analyte 2000™ biosensor is affordable and can be a sensitive and specific pathogen detection device. The project team examined the use of this biosensor to detect pathogens using the highly specific interactions between deoxyribonucleic acid (DNA) sequences. The aim was to develop an assay that could detect *C. parvum* and *H. pylori* at low levels by detecting genomic DNA sequences.

However, the team encountered major problems associated with using the biosensor in this fashion. While initial results appeared promising, subsequent experimentation determined that this assay approach was problematic. Consistently, the team achieved results that indicated no target DNA was present in the system. Exhaustive studies were performed to ensure that detection of the target DNA was specific to the system and that all DNA sequences used to both capture and detect DNA were not interacting in a non-specific manner. In addition, procedures were developed to amplify any signal generated. This also caused problems since the signal: background ratio was significantly high, independent of the presence of target DNA. These problems were not overcome through this research. Thus, the Analyte 2000™ biosensor cannot be used at this time to detect the DNA of pathogens of concern to the water/wastewater industries.

To complement these biosensor studies, the team investigated the use of polymerase chain reaction (PCR) and real time PCR as mechanisms to detect pathogens in well water and wastewater. Protocols were developed for *H. pylori,* and the team produced procedures that can detect low cell numbers. These methods can be used to detect <20 pg DNA, which is equivalent to 13 cells. However, this procedure requires extensive DNA purification and is not rapid. To speed up the process, we attempted to use real-time PCR of cells lysed directly from the water. While amplification was achieved, it was not of the correct sequence. This was overcome by improved cleaning processes and increased cell numbers.

Overall this project was successful in that detection procedures were developed using PCR. However, the major thrust of the project was to use the biosensor, and this proved to be unsuccessful. The project team studied every aspect of this method and concluded that using this biosensor to measure DNA:DNA interactions will not work.

CHAPTER 1.0

INTRODUCTION AND METHODS

The goals of detecting microorganisms using biosensors are to provide high sensitivity and specificity combined with a significant reduction in sample preparation, assay time, and the use of expensive reagents. Biosensors have been used to detect whole microbial cells, toxins, and nucleic acids. This detection can be performed using different techniques and various types of biosensors. Piezoelectric mass-sensing biosensors detect the presence of molecules using quartz crystal microbalance (QCM) resonators that measure resonance frequency changes that occur when a binding event takes place at the surface of the sensor (Zhou et al., 2001). Surface plasma resonance biosensors are similar to quartz crystal microbalances in that they also detect changes in resonance. However, this system utilizes light energy to sense changes in refractive indexes and alterations in the dielectric current (Chen et al., 1998). A biosensor developed by Mascini et al. (2000) measures deoxyribonucleic acid (DNA) hybridization as the double-strand formation occurs on its electrode. The hybridization is detected as a change in voltage. Nucleic acid detection has also been developed using fiber-optic technology. Fiber-optic gene arrays were used by Steemers et al. (2000) to detect unlabeled target DNA with molecular beacons.

The Analyte 2000™ biosensor is a fiber-optic biosensor that uses polystyrene tapered waveguides, to which targets with fluorescent molecules can be bound. Once fluorescent molecules are excited at this wavelength, the emissions can be detected between the range of 670 nm and 710 nm (Templeman et al., 1996). These wavelengths were chosen because fluorescence from naturally occurring molecules is limited (Wadkins et al., 1995). A fluorescent molecule that meets the requirements of this biosensor system is cyanine-5 (Cy-5).

The Analyte 2000™ biosensor, a flow through injection system, has been used to detect between 3 CFU/mL to 30 CFU/mL of *Escherichia coli* O157:H7 in ground beef (DeMarco et al., 1999). The bacteria were detected through a sandwich assay in which a captured antibody bound to the polystyrene waveguide immobilizes the target. The detection antibody, which is labeled with Cy-5, is then introduced, providing the fluorescent molecule necessary to generate a signal.

In theory, target DNA molecules could be detected in a similar way. A capture probe could be constructed to bind to a target gene and then detected with a detection probe labeled with Cy-5. Biotinylated capture probes can readily bind to streptavidin-coated surfaces, such as a waveguide or an enzyme-linked immunosorbent assay (ELISA) plate. Other DNA-binding methods have been explored, such as the use of amino modified oligonucleotides to bind to a carboxylate modified surface, or by binding DNA to a surface by dry-adsorption (Hirayama et al., 1996, Nikiforov and Rogers, 1995).

This project was comprised of two major components to identify a rapid and sensitive method to detect two waterborne pathogens: *Cryptosporidium parvum* and *Helicobacter pylori*. Both components utilize detection of specific components of the pathogens' DNA. First, the project team evaluated the use the Analyte 2000™ biosensor to detect specific DNA sequences. This was done by identifying specific gene targets for pathogen detection, synthesizing the necessary DNA reagents for use with the biosensor, followed by an exhaustive evaluation of the potential of the system. An amplification system was used to increase sensitivity of the biosensor. Second, the project team used the polymerase chain reaction (PCR) as a detection system. Both conventional PCR and real-time PCR was evaluated in its efficacy for detection of *H. pylori* cells in well water. This allowed for a direct comparison of sensitivities, ease of processing, specificity, and costs.

1.1 Target Genes for Detection

Target genes to be used for detection of *H. pylori* and *C. parvum* were identified through literature and database searches. It is important to use purified target DNA for initial analysis and characterization of the Analyte 2000™ biosensor as a DNA biosensor. For this reason target genes were identified that could be amplified by PCR. The amplicons will be used for DNA hybridization studies. The target genes and primers used were either designed or chosen from the literature and tested for specificity by Basic Local Alignment Search Tool (BLAST). At least two genes and their respective primer sets were chosen for each organism. This was to increase the possibilities of efficient PCR and increased specificity and DNA hybridization during the detection.

1.2 Cell Culture and DNA Extraction from *H. pylori*

Bacterial strain and culture conditions. The project team used *H. pylori* ATCC strains 43504 and 700392. A 100-μL aliquot from frozen stock that contained approximately 10^8 colony forming units (CFU)/mL was spread-plated onto trypticase soy agar with 5% sheep blood (BBL Prepared Media) that had been amended with antibiotics to contain vancomycin, amphotericin B, and trimethoprim. The antibiotics were added to plates by spreading 100-μL aliquots of each stock antibiotic solution: vancomycin (2 mg/mL), amphotericin B (2 mg/mL), and trimethoprim (1 mg/mL). The antibiotics were absorbed into the agar, giving a final concentration of approximately 10 μg/mL for vancomycin and amphotericin B and 5 μg/mL for trimethoprim. The plates were incubated microaerophilically using the BBL GasPak™ Jar System (Becton Dickinson, Franklin Lakes, NJ, USA) for 5 days at 37°C.

Harvesting of cells. Two milliliters of beef heart infusion (BHI) broth was pipetted onto each plate. A sterile glass rod was gently rubbed around the plate to re-suspend the bacterial colonies. The suspended culture was then pipetted into a sterile conical tube.

Extraction of DNA. Two milliliters of the suspended culture were used for DNA extraction with a commercial kit (DNeasy™ Tissue Kit, Qiagen Inc., Valencia, California). The manufacturer's instructions were followed, except that the final elution step was modified by eluting twice with 75 μl of elution buffer. The DNA was quantified using a Beckman DU640 Spectrophotometer (Beckman Coulter, Fullerton, California).

The absorbance (A) was measured at 260 nm, and the purity of the sample was determined by the ratio of A_{260}/A_{280}.

1.3 PCR Synthesis of Specific DNA Targets to Use as a Target for Detection Analysis of *H. pylori* on the Analyte 2000™ Biosensor

PCR amplification of *H. pylori* target DNA was attempted using a protocol taken from Bickley et al. (1993), author of the *glmM* (formerly *ure*C) primers. However, the desired results were not obtained, so a generic PCR protocol was taken from the HotStarTaq™ handbook (Qiagen, Valencia, California). Tables 1-1 and 1-2 provide the components for the synthesis of the *glmM* gene by PCR and the protocol for synthesis of that gene, respectively. Originally the team used a region of the *Urease* A gene. However, it was decided that smaller target DNA might be easier to detect with the biosensor. The size of the *H. pylori glmM* amplicon is 294 bp and the size of the *ure*A amplicon is 411 bp. The team originally was leaning toward using the *ure*A amplicon because preliminary PCR reactions of the *glmM* amplicon appeared to have an extraneous band present. Since then, all other PCR reactions (at least five) have shown only one concise band for the *glmM* amplicon at 294 bp. Also, when designing both capture and detection probes for both sequences it was necessary to design probes that had specific spacing parameters and also had very low self-dimerizing and hairpin formation properties. When designing probes for both *glmM* and *ureA* amplicons, the *glmM* sequence was much more compatible with parameters required for probe design. The criteria for spacing and compatibility parameters included: probe sequences could not be at the ends of the gene; capture and detection probes had to target sequences at least 30 bp apart to prevent steric hindrance; guanine and cytosine (GC) content had to be close to 50%, low self-dimerization, low inter-dimerization, and limited hairpin formation were necessary.

Table 1-1. Components for Synthesis of the *glmM* Gene by PCR.

Reagent	Concentration
glmM forward primer	0.5 mM
glmM reverse primer	0.5 mM
Taq (HotStarTaq)	1.5 units
PCR buffer	1X
dNTP	0.25 mM each

Table 1-2. PCR Protocol for Synthesis of the *glmM* Gene.

Cycle	Temperature	Time	Repeats
1	95°C	15 min	0
	95°C	1 min	
2	55°C	1 min	34
	72°C	1 min	
3	72°C	15 min	0

1.4 Cell Culture and DNA Extraction from *C. parvum*

Pathogen Strain: A sample of viable *C. parvum* oocysts at a concentration of 10^7/mL was obtained from a controlled diet bovine fecal samples (source of oocysts: Marilyn Marshall, University of Arizona, Department of Veterinary Science/Microbiology, Tucson, Arizona).

Oocyst lysis: A 1 mL sample of 10^6 oocysts was centrifuged at 5000 x g for 3 minutes to pellet oocysts. The pellet was then resuspended in 500 µL of 1X PCR buffer with 15 mM $MgCl_2$. The suspension was then exposed to freeze-thaw treatment to rupture oocysts and release nucleic acid. This involved seven cycles of freezing samples in liquid nitrogen for 1 minute and then thawing at 65°C for 1 minute. A final 30-minute boiling step was performed to precipitate and/or destroy contamination from host fecal matter. This involved three 10-minute boiling periods with a short pulse vortex in between periods. The lysate was spun at 16,000 x g for 5 minutes and 50 µL was removed from the meniscus to capture DNA.

C. parvum DNA isolation and purification: Samples were treated using a QIAamp™ DNA mini kit (Qiagen, Valencia, California) following the protocol for crude cell lysates as follows:

1. 150 µL of lysis buffer was added to the 50 µL lysate sample.
2. 20 µL of proteinase K was added.
3. 200 µL of buffer AL was added.
4. The solution was mixed immediately by pulse-vortexing for 15 seconds.
5. The solution was incubated at 56°C for 10 minutes and centrifuged briefly.
6. 200 µL of 100% ethanol was added and mixed again by pulse-vortexing for 15 seconds then followed by centrifugation.
7. 620 µL of lysate was added to the QIAamp spin column and centrifuged at 6000 x g (8000 rpm) for 1 minute. Filtrate was discarded.
8. 500 µL of buffer AW1 was added to the spin column. Column was then spun at 6000 x g (8000 rpm) for 1 minute. Filtrate was discarded.
9. 500 µL of buffer AW2 was added to the spin column. Column then was spun at full speed (20,000 x g; 14,000 rpm) for 3 minutes.
10. 200 µL buffer AE was added and allowed to incubate at room temperature for 1 minute. Column was centrifuged at 6000 x g (8000 rpm) for 1 minute and filtrate was collected and stored at -20°C.

1.5 PCR Synthesis of Specific DNA Targets to Use as a Target for Detection Analysis for *C. parvum* on the Analyte 2000™ Biosensor

DNA representing the target gene, *hsp70*, was amplified for use in analysis and standardization of the Analyte 2000™ biosensor as a DNA biosensor. PCR of the *C. parvum* genomic DNA was performed using the primer pairs previously described (Stinear et al., 1996), and the PCR cycling protocol was taken from the HotStarTaq™ handbook. Tables 1-3 and 1-4 provide components for synthesis by PCR and the protocol, respectively.

Table 1-3. Components for Synthesis of the *hsp70* Gene by PCR.

Reagent	Concentration
Chsp1 forward primer	2.5 mM
Chsp2 reverse primer	2.5 mM
Taq (HotStarTaq)	1 unit
PCR buffer	1X
dNTP	0.2 mM each

Table 1-4. PCR Protocol for Synthesis of the *hsp70* Gene.

Cycle	Temperature	Time	Repeats
1	95°C	15 min	0
2	95°C	1 min	39
	55°C	1 min	
	72°C	1 min	
3	72°C	10 min	0

1.6 Primer Design and Testing for Target Capture and Detection of *H. pylori* and *C. parvum* DNA on the Analyte 2000™ Biosensor

Biosensor assays for detection of the target DNA were originally designed to make use of the Biotin Tyramide System (BTS) amplification system to detect purified PCR amplicons. This system, described in Section 1.8, requires an antigenic tag on the detection probe.

For *H. pylori,* the target amplicon is a 294-bp region of the phosphoglucosamine mutase *glmM* gene (old *Urease C*) previously described in Section 1.3. In order for the capture probe to be linked to the waveguide, the waveguide is first coated with streptavidin to bind the biotin on the capture probe. The capture probe (5'-CTC ACA AAC TTA TCC CCA ATC GC-3') was 5' end labeled with biotin and binds to a region within the *glmM* gene, the detection probe (5'- TAT GCC CGC TTT GCT GCG CTC CAA-3') was 5' end labeled with fluorescein isothio-cyanate (FITC) and also binds to a region within the *glmM* gene.

For *C. parvum* the target PCR amplicon was a 590 bp region of the *hsp70* gene from *C. parvum*. The capture probe (5'-CCA TAG AAC TTA CAC AGA CGA C-3') was 5' end labeled with biotin and binds to a region within the *hsp70* gene, the detection probe (5'-GAC CAT TGT CGT TCT AGT GAT C-3') was 5' end labeled with FITC and also binds to a region within the *hsp70* gene.

1.7 Analysis of Binding of Capture Probes to Waveguides

It was important to achieve optimal binding of capture probes to the waveguides since this ultimately would result in more efficient binding of the target DNA and detection probes to the system. To analyze the binding of the capture probes via the biotin: streptavidin interaction, the team utilized a radioactively labeled capture probe and analyzed how much probe was binding and the location of the binding on the waveguide.

A radioisotope-labeled capture probe was allowed to bind to streptavidin-coated waveguides. One of the waveguides was placed as one whole piece into a scintillation vial. The other waveguide was cut into 11 sections, and each section was placed in scintillation fluid. These sections were read using a scintillation counter. Since the waveguide is kept in a chamber and reagents flow from one end of the chamber (proximal end) to the distal end, it was worth noting where most of the binding occurred. Signal received from the scintillation counter was measured as counts per minute (cpm).

1.8 Evaluation of the BTS

The BTS is a method of signal amplification used both in Analyte 2000™ biosensor and ELISA reader fluorescent assays (Figure 1-1). The binding surface (either waveguide or micro titer plate) was first coated overnight with streptavidin. Next, a biotin labeled capture probe, a 5' end labeled FITC detection probe, and target amplicon were allowed to bind to the coated surface. The FITC is used as a target antigen so that horseradish peroxidase can be specifically located into the system. The FITC is linked to the detection probe, which has a complementary sequence to the target DNA. This means that should hybridization of the target DNA to the capture probe and the detection probe occur, then the detection probe is immobilized onto the waveguide or ELISA plate surface. The waveguide was blocked to prevent non-specific binding and then an anti-fluorescein horseradish peroxidase (HRP) conjugated antibody was applied to the system. Biotin tyramide was added next, as this reacts with the HRP. Reactive biotin tyramide deposits on and around the vicinity of HRP and is responsible for the amplification. This is because the biotin tyramide supplies a target for capture of the Cy-5 fluorescence molecule. Cy-5 conjugated streptavidin was applied to the waveguide and will bind to any available bound biotin tyramide. The signal given off by the bound Cy-5 was read either in picoamps (Analyte 2000™ biosensor) or relative fluorescent units (ELISA reader). The intensity of the signal should be related to the amount of target DNA captured by the capture probe.

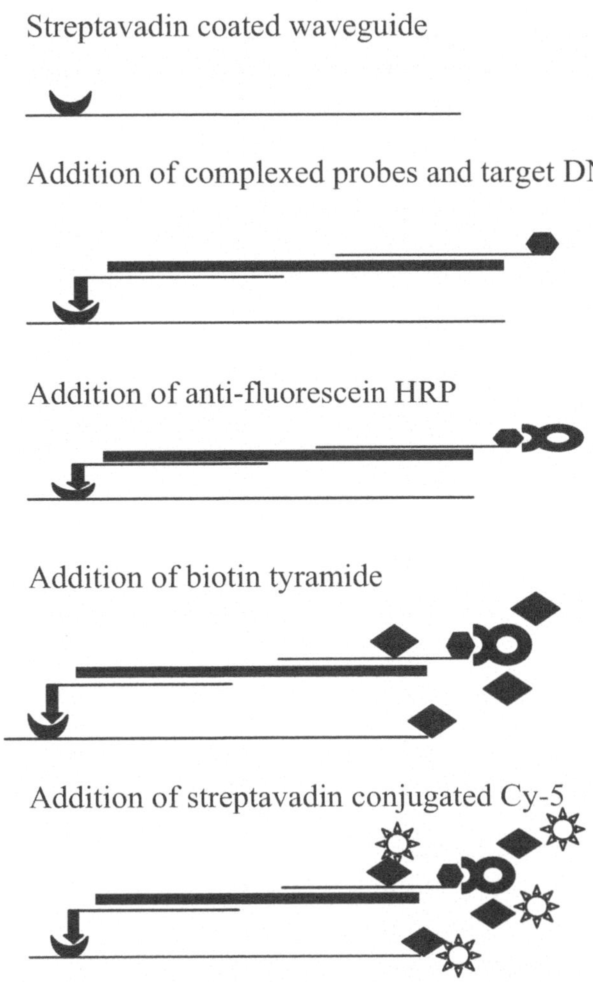

Figure 1-1. Diagram of the Use of the BTS in the Detection of DNA Targets.

All experiments that included the BTS system were performed using the following protocol:

The binding surface (either waveguide or micro-titer plate) was incubated at 4°C overnight with streptavidin (100 µg/mL in phospated buffer saline [PBS]). The binding surface was washed three times with 1X PBST (phosphate buffered saline, 0.1% Tween 20). Next, the binding surface was bathed with 2X standard sodium citrate, pH 7.0 (SSC). Probes (capture and detection probes specific for the target DNA) and target DNA were diluted from the stock solutions to the appropriate concentrations in 2X SSC. Capture, detection, and target DNA were combined, then heated to 95°C for 10 minutes and placed on ice prior to exposure to the binding surface. The DNA mixture (probes and target) was allowed to incubate for 1 hour at room temperature. Following incubation, the binding surface was washed with 2X SSC. The surface was then blocked using 2% bovine serum albumin (BSA) and 2% casein. The blocking buffer was allowed to incubate at room temperature for 30 minutes. Next, 5X anti-fluorescein HRP diluted in blocking buffer was applied and allowed to incubate at room temperature for 1 hour. The surface was then washed with 1X PBST. Next, 20 μL biotin tyramide was diluted in 500 μL water, and 500 μL diluent provided by the manufacturer was added and allowed to incubate at room temperature for 10 minutes. The surface was washed with 1X PBST, and the initial reading was taken. Cy-5-labeled streptavidin was diluted in blocking buffer to a concentration of 1 μg/mL and was incubated for 10 minutes at room temperature. The surface was washed with 1X PBST then with 2X SSC and a reading was taken.

A summary of this protocol is given here:

1. Wash 1X PBST
2. Wash 2X SSC (take reading)
3. DNA 1-hour at room temperature
4. Wash 2X SSC
5. Blocking buffer
6. Anti-fluorescein HRP
7. Wash 1X PBST
8. Biotin Tyramide
9. Wash 1X PBST
10. Cy-5 streptavidin
11. Wash 1X PBST
12. Read in 2X SSC

1.9 Direct Fluorescent Labeling Detection System

To further determine the effectiveness of the BTS system, direct fluorescent detection assays were performed on both the Analyte 2000™ biosensor and the ELISA fluorescent reader. In this case, the detection probe used for detection of the target was labeled with a fluorescent signal, and there was no further enhancement of the signal by amplification with BTS. The binding surface (either waveguide or micro-titer plate) was first coated with streptavidin. Next, capture probe (biotin labeled), a detection probe (FITC or Cy-5 labeled), and target amplicon were allowed to bind to the streptavidin-coated surface. The signal was read in either picoamps (biosensor) or relative fluorescent unit (ELISA reader).

The general method for the direct labeling detection system was as follows:

The binding surface (either waveguide or micro-titer plate) was incubated at 4°C overnight with streptavidin (100 μg/mL in PBS). The surface was washed with 1X PBST then washed with 1 mL 2X SSC. Probes and target DNA were diluted from the stock solutions to the appropriate concentrations in 2X SSC. An initial reading was taken to obtain the pre-detection signal. Capture probe, detection probe, and target DNA were combined, heated to 95°C for 10 minutes, and placed on ice until exposure to the binding surface. The non-target DNA controls contained capture and detection probes only. The DNA mixture was allowed to incubate for 1 hour at room temperature. Following incubation, the binding surface was washed with 1 mL 2X SSC, and the final reading was taken.

A summary of this protocol is given here.

1. Wash 1X PBST
2. Wash 2X SSC (take reading)
3. DNA 1-hour at room temperature
4. Wash 2X SSC
5. Wash 1X PBST
6. Read in 2X SSC

1.10 PCR-Dependent Labeled Amplicon Detection System

PCR-dependent labeled amplicon assays were performed on both the Analyte 2000™ biosensor and the ELISA fluorescent reader. The binding surface (either waveguide or micro-titer plate) was first coated with streptavidin. Internally labeled Cy-5 amplicon was synthesized by PCR using Cy-5 labeled dCTP and then applied to the streptavidin coated binding surface with the biotin labeled capture probe. The signal was read in either picoamps (biosensor) or relative fluorescent unit (ELISA reader).

The binding surface was incubated at 4°C overnight with streptavidin (100 μg/mL in PBS). The surface was washed with 1X PBST then washed with 1ml 2X SSC. Biotin-labeled capture probe and Cy-5-labeled target amplicon were diluted from the stock solutions to the appropriate concentrations in 2X SSC, and an initial reading was taken. Capture probe and target DNA were combined, then heated together to 95°C for 10 minutes, and then placed on ice until exposure to the binding surface. The DNA mixture was allowed to incubate for 1 hour at room temperature. Following incubation, the binding surface was washed with 1 mL 2X SSC, and the final reading was taken.

A summary of this protocol is given here.

1. Wash 1X PBST
2. Wash 2X SSC (take reading)
3. DNA one hour RT
4. Wash 2X SSC
5. Wash 1X PBST
6. Read in 2X SSC

1.11 Immunomagnetic Separation and Filtration of *H. pylori* Cells

Immunomagnetic separation (IMS) was used as a method to both concentrate and separate the *H. pylori* cells from well water. Dynal Dynabeads M-280™ (Dynal, New Hyde Park, New York) are magnetic beads pre-coated with sheep anti-rabbit immunoglobin G (IgG). They were incubated at 4°C for 24 hours with polyclonal rabbit anti-*H. pylori* IgG (Biomeda, Foster City, California). Next, 2 µL of antibodies (1.4 mg/mL) were used per 10 µL of paramagnetic beads. The beads were washed gently with PBS/1% fetal bovine serum (FBS) before and after incubation with the antibodies. A magnet was used to draw the beads to one side of the microcentrifuge tube so that the wash could be drawn off.

Water samples inoculated with *H. pylori* were concentrated by filtration onto 0.2 µm Supor membranes (Pall Corporation, Ann Arbor, Michigan). The filters were transferred to 15 mL Dynal tubes that contained 5 mL of PBS/1% FBS. Twenty microliters of complexed beads (beads plus primary antibody) were added per milliliter giving a final concentration of 1.3×10^7 beads/mL of concentrated sample. After the tubes were gently rotated for 1 hour at 4°C, the samples were washed twice with PBS/1% FBS. The beads with the targeted *H. pylori* were then resuspended in 50 µL to 100 µL of sterile water and boiled for 10 minutes to lyse the cells. The actual recovery of cells using the method was not determined.

1.12 Conventional PCR for Detection of *H. pylori* Cells

The T-personal thermocycler (Whatman Biometra, Göttingen, Germany) was used for conventional PCR. Each reaction contained: 5 µL 10X Buffer B and 0.25 µl Taq DNA polymerase (5 U/µL) (Fisher, Pittsburgh, Pennsylvania); 1 µL dNTP 10 mM (Promega, Madison, Wisconsin); and a final concentration of 2 mM $MgCl_2$ and 0.5 µM of each primer. The previously published primers targeting the 16S ribosomal ribonucleic acid (rRNA) gene (Engstrand et al., 1992) and the urease gene, subunit A (Clayton et al., 1992) were purchased from IDT (Coralville, Iowa). The thermocycler program for targeting the 16S rRNA (Engstrand et al., 1992) was modified, changing the extension step to 1 minute, as seen in Table 1-5. For targeting the *ureA* gene (Clayton et al., 1992), the denaturation step was modified to 94°C for 10 minutes, and final elongation was extended to 10 minutes, also seen in Table 1-5.

Table 1-5. Protocol for Detection of *H. pylori* by PCR.

Primers for:	Denaturation	Amplification
16S rRNA gene	94°C 5 minutes	39 cycles: 94°C 1 minute 55°C 1 minute 72°C 1 minute, final elongation 10 minutes
ureA gene	94°C 10 minutes	35 cycles: 94°C 1 minute 45°C 1 minute 72°C 1 minute, final elongation 10 minutes

1.13 Real time PCR Detection of *H. pylori*

Real Time PCR was used as a rapid method to detect *H. pylori* cells in well water. Real Time PCR was performed with the Lightcycler and FastStart™ DNA Master SYBR Green I kit (Roche, Indianapolis, Indiana) using the conditions described in Table 1-6. Primers to the 16S rRNA and the *ureA* gene targets were used for detection. In some cases real-time PCR agarose gels were used to analyze products produced from real-time PCR. All agarose gels were 1% (weight per volume) and electrophoresis was performed at 95V for 1 hour.

Table 1-6. Conditions for Real Time PCR of *H. pylori* Genes.

Primers for:	Denaturation	Amplification	Melting Curve
16S rRNA gene	95°C 10 minutes	45 cycles: 95°C 15 seconds 55°C 5 seconds 72°C 25 seconds	95°C 0 seconds 65°C 15 seconds 95°C 0 seconds
ureA gene	95°C 10 minutes	45 cycles: 94°C 15 seconds 45°C 5 seconds 72°C 21 seconds	95°C 0 seconds 65°C 15 seconds 95°C 0 seconds

CHAPTER 2.0

RESULTS

2.1 Target Genes for Detection

The target genes to be used for detection of *H. pylori* and *C. parvum* were identified by literature and database search. The *ureA* and *glmM* genes of *H. pylori*, and the heat shock protein (*hsp70*) of *C. parvum* were targeted with specific primers. A 458-base-pair fragment amplified with random primers was also utilized. At least two genes and their respective primer sets were chosen for each organism in order to increase the possibilities of success. Below, the target genes for each organism are given with the forward and reverse PCR primers.

H. pylori
*ure*A gene:
The primers shown below would amplify a 411 bp fragment, nucleotides 304-714 of the *ureA* gene (Clayton et al., 1992).

HPU1, 5'-GCCAATGGTAAATTAGTT-3'
HPU2, 5'-CTCCTTAATTGTTTTTAC-3'

*ure*C *gene* (now known as *glm*M gene):
The primers listed below would amplify a 294 bp fragment, nucleotides 784-1077, (Bickley et al., 1993).

*glm*M F, 5'-AAGCTTTTAGGGGTGTTAGGGGTTT-3'
*glm*M R, 5'-AAGCTTACTTTCTAACACTAACGC-3'

16S rRNA gene:
The primers below would amplify a 522 bp fragment, nucleotide 219-740, (Engstrand et al., 1992).

Hp1, 5' – TGGCAATCAGCGTCAGGTAATG – 3'
Hp2, 5' – GCTAAGAGATCAGCCTATGTCC – 3'

C. parvum
hsp70 gene Heat Shock Protein 70 (Chsp):
The primers below would amplify a 590 bp fragment, nucleotides 2431-3020, (Stinear et. al. 1996)

Chsp1, 5'-AGCAATCCTCTGCCGTACAGG-3'
Chsp4, 5'-AAGAGCATCCTTGATCTTCT-3'

C. parvum- SBO12 (random amplified polymorphic DNA (RAPD):
The primers below would amplify a 458 bp fragment (Wu et al., 2000).

SBO12F, 5'-ATAAACAAGCAGGAAAAAAGGT-3'
SBO12R, 5'-GCGACAAGTTACAAGGATTATT-3'

Before using the above primers, their homology to other sequences in GenBank (www.ncbi.nlm.nih.gov/entrez) was checked with via BLAST. The primers were homologous only to their respective target gene sequences. The potential for primer dimer formation was also found to be minimal. No mismatches were allowed for homology between primers and their target sequences. The potential for primer:dimer formation was determined using alignment software available at the IDT Web site.

2.2 Cell Culture and DNA Extraction from *H. pylori*

DNA was extracted from *H. pylori* cells using the method described. Stock 1 and Stock 2 refer to separate extractions of DNA from different *H. pylori* cultures. Analysis of the extracted DNA by agarose gel electrophoresis (Figure 2-1) determined that the DNA appeared to be somewhat fragmented as shown by the smear on the gels. However, it was decided that for this application the DNA would likely be an adequate template for PCR amplification of the target amplicons.

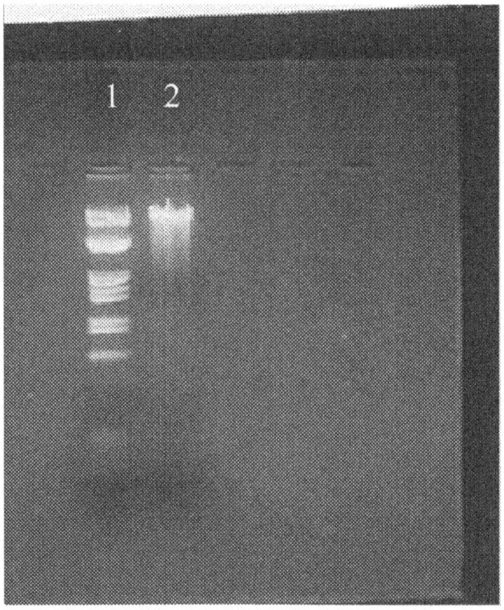

Figure 2-1. Undigested, Genomic DNA Extracted from *H. pylori*. Lane 1, Molecular Weight Marker; Lane 2, *H. pylori* Stock 1.

2.3 PCR Synthesis of Specific DNA Targets to Use as a Target for Detection Analysis of *H. pylori* on the Analyte 2000™ Biosensor

DNA representing the target genes *ureA* and *ureC* was amplified for intended use in analysis and standardization of the Analyte 2000™ biosensor as a DNA biosensor. PCR of the *H. pylori* genomic DNA extracted as in the previous section was performed under standard conditions using the primer pairs previously described for *ureA* and *ureC*. The amplification was performed using the conditions described in the methods section. The amplified products were visualized via agarose gel electrophoresis. The gel for the *ureC* gene, now called *glmM* is shown in Figure 2-2.

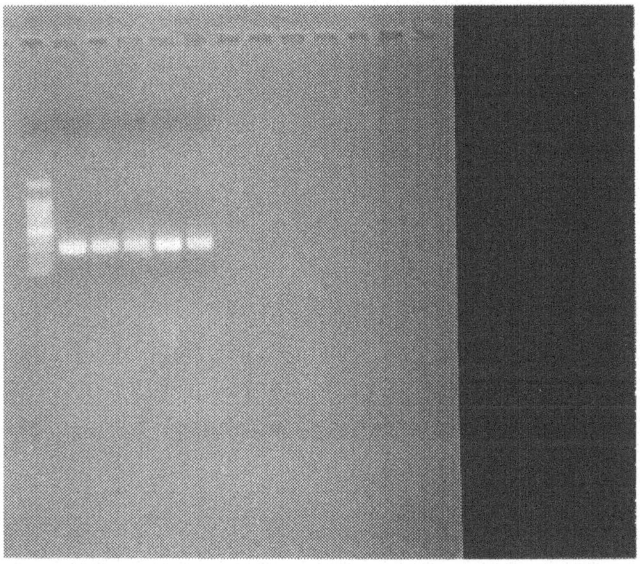

Figure 2-2. *glmM* Amplicons from *H. pylori*. Lane 1, Molecular Weight Marker; Lanes 2 – 6, *glmM* Target from 1 μL Template.

The *glmM* products were purified for use as target DNA on the Analyte 2000™ biosensor. These amplicons were also verified by DNA sequencing (Figure 2-3) prior to further use. Sequencing was performed at a University of South Florida laboratory with sequencing facilities. The amplified sequences were nearly 100% homologous with the gene sequences in Genbank. There were a few bases that were difficult to read (N), but discounting these, there was 100% identity. The amplicon showed 100% identity at the sites of capture probe and target probe binding. For reasons of PCR purity and size the *glmM* amplicon was the target used for all future analysis.

5'-TCTNNGGTANNNNNCTCTGGNANCNGCTCCNAGAGCGAATTAGTNCGCTACGAG

CATGAGCAATGTTAGCCCTTAAAGAATACCTAAAATNCCCAAGATNTTAGAATTGA

capture sequence
AGCATGNGC***GCGATTGGGGATAAGTTTGTGAG***CGAATGCANTGCGATTGAACAAAG

detection sequence
CCAATNT***TTGGAGCGCAGCAAAGCGGGCATA***TCATTTTATANGCGATTACGCNTAAA

ACCGGCGATGGTTNNGGTGTGCGCTTTGCAAGTGAGTGCGTNAGTGTTAGAAAGTA

AGCNNANA-3'

Capture Probe

```
                   CTC ACA AAC TTA TCC CCA ATC GC -3'
                   ||| ||| ||| ||| ||| ||| |||
Sequence data 3'-GAG TGT TTG AAT TGG GGT TAG CG -5'
```

Detection Probe

```
                   TAT GCC CGC TTT GCT GCG CTC CAA -3'
 ||| ||| ||| ||| ||| ||| ||| |||
Sequence data 3'-ATA CGC GCG AAA CGA CGC GAG GTT- 5'
```

Figure 2-3. Sequencing Data Generated from Analysis of *H. pylori*, *glmM* Target Amplicon Using the Forward Primer. Capture and Target Sequences are the Sequences Used Later On for the Detection in the Analyte 2000™ Biosensor. These Sequences Were Maintained in the Amplicon.

2.4 Cell Culture and DNA Extraction from *C. parvum*

DNA was extracted from *C. parvum* oocysts using the described method. The DNA was of adequate quality for use as a template for PCR amplification of the target amplicons (results not shown).

2.5 PCR Synthesis of Specific DNA Targets to Use a Target for Detection Analysis for *C. parvum* on the Analyte 2000™ Biosensor

The *hsp70* gene for the *C. parvum* heat shock protein 70 was the target of PCR. DNA representing the target gene for *hsp70* was amplified for intended use in analysis and standardization of the Analyte 2000™ as a DNA biosensor. PCR of the *C. parvum* genomic DNA extracted as in the previous section was performed under standard conditions using the primer pairs previously described. The amplification was performed using the conditions described in the methods section. The amplified products were visualized via agarose gel electrophoresis. The gel for the amplification of the *hsp70* gene is shown in Figure 2-4.

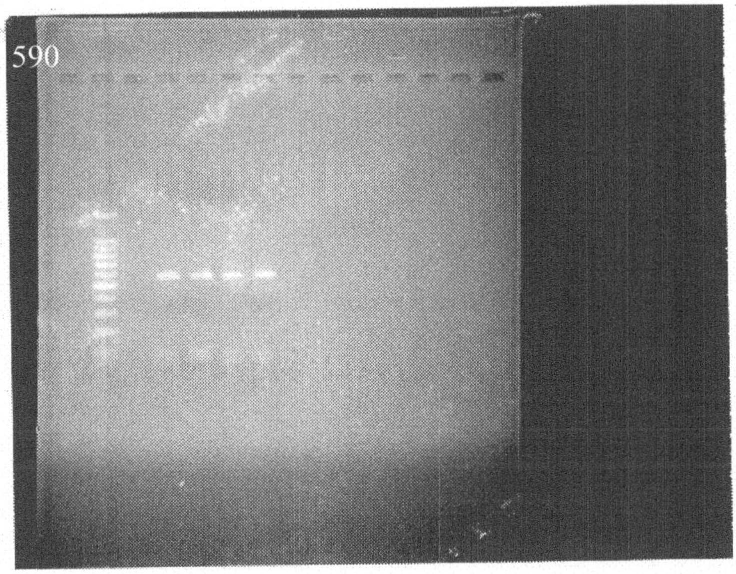

Figure 2-4. *Hsp70* Amplification from *C. parvum* by PCR. Lane 1, Molecular Weight Marker; Lanes 2 –5, Amplicon from 5 μl Template (Extracted from 10^6 Oocysts).

These amplified products were then purified for use as target DNA on the Analyte 2000™ biosensor. These amplicons were also sequenced prior to use to verify they were the correct genes amplified (Figure 2-5). The amplified sequences were homologous with the gene sequences in Genbank. There were a few bases that were difficult to read (N), but discounting these, there was 100% similarity. The amplicon showed 100% identity at the sites of capture probe and target probe binding. These results show that the sequences we are amplifying for target detection by the biosensor are indeed the correct sequences.

5'-ATGNGATGTNGCTCCATTATGCACTCGTTTAGAAACGCTGGTGGTGTGATGACCA

AGNCTTATTGAACGTAATACNACTATCCCCAGCAAAGAAGACACAAGTCTTCACTAC

capture sequence

TTATGCTGATAACCAGAGTGGTGTCTGATCCAAG***TTTATGAGGGTGAGAGAGCCA***

detection sequence

TGACTAAGGATAACCATCTCCTCGGAAAGTTCCATCTTGATGG***TATTCCACCAGCA***

CCAAGAGGTGTACCACAAATTGAAGTCACCTTTTGATATNGATGCTAATGGTATCTT

GAATGTGTCTGCTGTTGATAAGAGTACTGGTAAGAGCAGCAAGATCACTATTACTAA

CGATAAGGGTAGATTATCAAAGGACGATATTGAACGTATGGTTAATGATGCTGAGA

AATACAAGGGTGAGGATGAGCAGAACAGACTTAAGATTT- 3'

Capture Probe

```
                  TGG CTC TCT CAC CCT CAT AAA -3'
                  ||| ||| ||| ||| ||| ||| |||
Sequence data 3'-ACC GAG AGA GTG GGA GTA TTT-5'
```

Detection Probe

```
                  ACC TCT TGG TGC TGG TGG AAT A -3'
              ||| ||| ||| ||| ||| ||| ||| |
Sequence data 3'- TGG AGA ACC ACG ACC ACC TTA T- 5'
```

Figure 2-5. Sequencing Data Generated from Analysis of the *C. parvum hsp70* Gene Amplicon Using the Forward Primer. Capture and Target Sequences are the Sequences Used Later On for the Detection on the Analyte 2000™ Biosensor. These Sequences Were Maintained in the Amplicon.

2.6 Primer Design and Testing for Target Capture and Detection of *H. pylori* and *C. parvum* DNA on the Analyte 2000™ Biosensor

The probes to be used for both capture and detection of the *H. pylori* and *C. parvum* target amplicons were designed based on GC content and location in the amplicons. Figures 2-6 and 2-7, respectively, show where the binding sites for these specific probes are located in their respective target genes for *H. pylori* and *C. parvum*.

294pb portion (Ac.# AF405553)
5'AAGCTTTTAGGGGTGTTAGGGGTTTATCAAAAATCTAAAAACGCCCTTTCTT

CTCAAGCAATTGTCGCTACGAGCATGAGCAATTTAGCCCTTAAAGAATACTT

capture sequence

AAAATCCCAAGATTTAGAATTGAAGCATTGT***GCGATTGGGGATAAGTTTGT***

detection sequence

GAGCGAATGCATGCAATTGAATAAAGCCAATT***TTGGAGGCGAGCAAAGC***

GGGCATATCATTTTTAGCGATTACGCTAAAACAGGCGATGGTTTGGTGTGCG

CTTTGCAAGTGAGCGCGTTAGTGCAAGTAAG- 3'

Capture Probe (Bio = biotin)
5' -/5Bio/CTC ACA AAC TTA TCC CCA ATC GC-3'   22bp; CG content 47.8%

Detection Probe (Fluor = fluoroscein)
5' -/5Fluor/TAT GCC CGC TTT GCT GCG CTC CAA-3'   24bp; CG content 58.3%

Figure 2-6. Probe Map of *H. pylori* phosphoglucosamine mutase (*glmM*) Gene to Identify the Sites of the Capture and Detection Sequences.

590 bp portion (ac.# U11761)

5'-
TTAAATGGTGAGCAATCCTCTGCCGTACAGGATCTCTTATTATTGGATGTTGC

TCCATTATCACTCGGTTTAGAAACTGCTGGTGGTGTTATGACCAAGCTTATTG

AACGTAATACAACTATCCCAGCAAAGAAGACACAAGTCTTCACTACTTATGC
capture sequence
TGATAACCAGAGTGGTGTCTTGATCCAAG***TTTATGAGGGTGAGAGAGCCA***
detection
TGACTAAGGATAACCATCTCCTCGGAAAGTTCCATCTTGATGG***TATTCCACC***
sequence
AGCACCAAGAGGTGTACCACAAATTGAAGTCACCTTTGATATTGATGCTAA

TGGTATCTTGAATGTGTCTGCTGTTGATAAGAGTACTGGTAACAGCAGCAAG

ATCACTATTACTAACGATAAGGGTAGATTATCAAAGGTATCTGATATTGAAC

GTATGGTTAATGATGCTGAGAAATACAAGGGTGAGGATGAGCAGAACAGAC

TTAAGATTGAGGCTAAGAACTCTTTGGAGAACTACCTCTATAACATGAGGAA

CACCATCCAAGAACCAAAGGTTAAGGAAAAGCTTTCTCAATCTGAAATTGAT

GAGGCTGAGAAGAAGATCAAGGATGCTCTT -3'

Capture Probe (Bio = biotin)
5' -/5Bio/TGG CTC TCT CAC CCT CAT AAA-3'   21bp; CG content 47.6%

Detection Probe (Fluor = fluoroscein)
5' -/5Fluor/ACC TCT TGG TGC TGG TGG AAT A-3'   22bp; CG content 50.0%

Figure 2-7. Probe Map of *C. parvum* Heat Shock Protein (*hsp70*) Gene to Identify the Sites of the Capture and Detection Sequences.

The possibility exists that non-specific annealing between the detection and capture probes could occur. Nonspecific annealing could result in a high background signal yielding false positives. One method to investigate this possibility is to examine the self and non-specific annealing properties of the probes using annealing software. Figure 2-8 shows the various binding characteristics of the probes used for capture and detection of both *H. pylori* (A) and *C. parvum* (B) DNA targets. In both cases there is some limited interaction between the probes. However, this would likely not occur under the hybridization conditions used in the actual assay. The project team also checked the self-annealing properties of the probes, as in binding to themselves. While this might not contribute to background, it could result in lower signal and sensitivity if it occurred. The analysis (not shown) suggested very little self annealing for each probe with itself.

A. *H. pylori*

```
1.                capture 5'-CTCACAAACTTATCCCCAATCGC-3'
                               |||| |         |
         detection 3'-AACCTCGCGTCGTTTCGCCCGTAT-5'

2.             capture 5'-CTCACAAACTTATCCCCAATCGC-3'
                          |   |||
                    detection 3'-AACCTCGCGTCGTTTCGCCCGTAT-5'

3.      capture 5'-CTCACAAACTTATCCCCAATCGC-3'
                                     |  || |
                      detection  3'-AACCTCGCGTCGTTTCGCCCGTAT-5'

4.                   capture 5'-CTCACAAACTTATCCCCAATCGC-3'
                             |  |   ||       |  |
             detection 3'-AACCTCGCGTCGTTTCGCCCGTAT-5'

5.              capture 5'-CTCACAAACTTATCCCCAATCGC-3'
                                          || |  ||
                    detection 3'-AACCTCGCGTCGTTTCGCCCGTAT-5'
```

B. *C. parvum*

```
1.               capture 5'-TGGCTCTCTCACCCTCATAAA- 3'
                         |   | |   ||| |   |
             detection 3'-ATAAGGTGGTCGTGGTTCTCCA- 5'

2.               capture 5'-TGGCTCTCTCACCCTCATAAA- 3'
     |  |  ||| |   |   ||
                    detection 3'-ATAAGGTGGTCGTGGTTCTCCA- 5'

3.               capture 5'-TGGCTCTCTCACCCTCATAAA- 3'
             ||   ||   |
        detection 3'-ATAAGGTGGTCGTGGTTCTCCA- 5'

4. capture 5'-TGGCTCTCTCACCCTCATAAA- 3'
       |    ||  |    |
            detection 3'-ATAAGGTGGTCGTGGTTCTCCA- 5'

5.               capture 5'- TGGCTCTCTCACCCTCATAAA- 3'
|    || |  ||   |
                 detection 3'-ATAAGGTGGTCGTGGTTCTCCA- 5'
```

Figure 2-8. Alignments that Yield the Highest Number of Complementary Base Pairs Between the Detection Probe and Capture Probe.

2.7 Analysis of Binding of Capture Probes to Waveguides

The labeled capture probe appeared to have a higher affinity to bind to the waveguide toward the proximal end (Figures 2-9 and 2-10). It is difficult to explain why this occurred. The labeled probe was introduced through the chamber first to the proximal end, and ultimately flowed out the distal end. An additional finding was the binding efficiency of the probe to the waveguide. The actual binding was considerably lower when compared to how much capture probe was washed out. Approximately 1% of the total amount of probe introduced was bound to the waveguide. The project team initially assumed that much of the remaining probe was binding to the chamber. They later confirmed that there was maximum binding capacity on the waveguides. The maximum capacity was found when the labeled capture probe was incubated with the waveguide in a glass capillary tube. Even though a larger amount of probe was used, nearly the same amount bound to the waveguide. This lack of binding efficiency is particularly problematic because it could potentially lead to the loss of target DNA using the sandwich type system. While hybridization may be occurring in solution, these hybrid molecules would get washed out of the system rather than bind to the waveguides. Consequently they would be undetectable. Binding the capture probe first to the waveguide followed by the introduction of the target could solve the problem. This binding, however, would cause an increase in assay time and could be less efficient given that the complementary strands of target would likely re-anneal and not bind to the shorter, but complementary strand of the capture probe. However, even though 100 femtomoles and 1000 femtomoles were added in various experiments, these dilutions were based on the estimate that 100% of the DNA was labeled with ^{32}P.

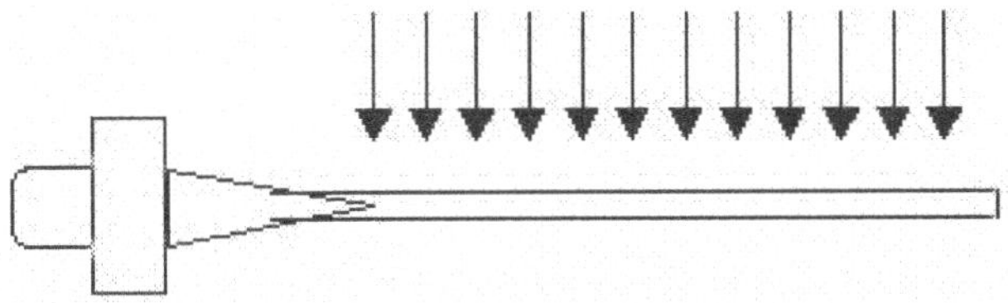

Figure 2-9. Location of Waveguide Separation for Radioisotope Studies. Arrows indicate points at which waveguide was divided into multiple fragments. The sections were placed separately in scintillation vials and measured using a scintillation counter to determine where along the waveguide binding was occurring.

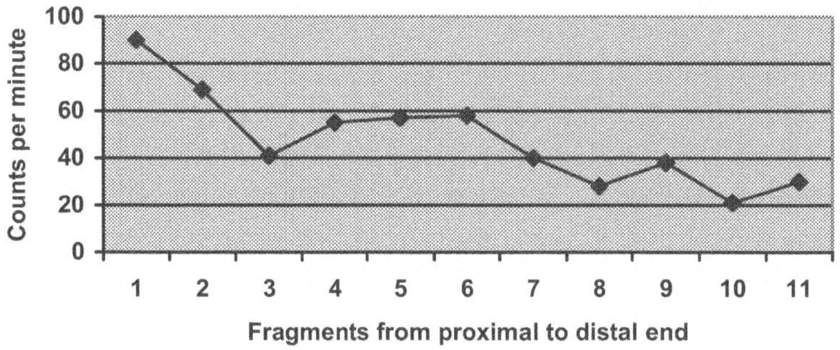

Figure 2-10. Distribution of Biotinylated Capture Probe Along Streptavidin-Coated Waveguide. Approximately 100 femtomoles of capture probe were incubated for 45 minutes within the chamber. The waveguide was cut into 11 Pieces (Figure 2-10) and measured using a scintillation counter. Blank readings of scintillation fluid are approximately 20 counts per minute. Results shown are for one experiment that is representative of two experiments that showed a similar trend. Differences in labeling resulted in differences in cpm, preventing averaging of the results.

2.8 Evaluation of the BTS

One major problem that was foreseen using the Analyte 2000™ biosensor for detection of pathogens via DNA:DNA interactions was the probability of low levels of molecules to detect, as in few gene numbers. For this reason it was decided to invest time in evaluating a system that would increase any signal generated as a result of DNA:DNA interactions. The BTS system was evaluated. This system (described in Section 1.8) uses horseradish peroxidase to increase biotin sites around the DNA:DNA interactions, allowing for the introduction of streptavidin tagged Cy-5 to increase the fluorescence signal. The following studies summarize many experiments to evaluate the efficacy of this system on the Analyte 2000™ biosensor. In most cases at least three trials were performed for each experiment. In every case, the experiments were performed with the *H. pylori* and *C. parvum* targets.

The initial experiments used the standard protocol detailed in the methods section. Table 2-1 shows the results of two trials where detection of *C. parvum hsp70* was attempted using the standard protocol with 0.5 and 5 pmoles of amplicon and 10 pmoles of each probe. The detection signals of amplicon were high, with abnormally high blank readings in all assays with little increase in signal as a result of presence of target DNA.

Table 2-1. Detection of *C. parvum hsp70* DNA Using the BTS System.

Trial 1	Non-Target Blank	0.5 pmoles Target	0.5 pmoles Target	5 pmoles Target
Initial signal	505	456	422	1102
Final signal	298	334	330	402
Δ Signal	-207	-122	-92	-700
Trial 2				
Initial signal	713	612	589	996
Final signal	419	397	412	496
Δ Signal	-294	-215	-177	-500

This was repeated for the *H. pylori glmM* gene using the standard protocol with 0.5 and 5 pmoles of amplicon and 10 pmoles of each probe (Table 2-2).

Table 2-2. Detection of *H. pylori glmM* Target Using the BTS System.

Trial 1	Non-Target Blank	0.5 pmoles Target	0.5 pmoles Target	5 pmoles Target
Initial signal	360	362	340	293
Final signal	417	373	391	355
Δ Signal	+57	+11	+51	+62
Trial 2				
Initial signal	341	645	558	499
Final signal	397	742	575	658
Δ Signal	+56	+97	+17	+159
Trial 3				
Initial signal	763	431	362	561
Final signal	727	386	480	587
Δ Signal	-36	-45	+118	+26

Again, little difference in the signal was observed irrespective of the presence of target DNA. Increasing the amount target DNA to 50 pmoles for both *H. pylori* and *C. parvum* targets did not resolve this (results not shown).

To determine if the BTS system was having any effect, experiments were designed whereby detection could be achieved using a detection probe labeled directly with Cy-5 with no BTS added to the system. However, a direct comparison could be made between the signals obtained with or without the addition of the BTS. The results of one set of trials for the *H. pylori* target are shown in Table 2-3. Here 50 pmoles of target DNA was used with 10 pmoles of capture and detection probe.

Table 2-3. Comparison of the BTS Amplification System with a Cy-5 Labeled Probe for Detection.

Trial 1	Cy-5 Blank	Cy-5 50pmoles Target	BTS Blank	BTS 50 pmoles Target
Initial signal	789	1094	557	406
Final signal	371	284	361	294
Δ Signal	-418	-810	-196	-112
Trial 2				
Initial signal	323	281	5703	1891
Final signal	7469	1418	5428	1732
Δ Signal	+7146	+1137	-275	-159
Trial 3				
Initial signal	320	338	477	331
Final signal	313	279	335	309
Δ Signal	-7	-59	-142	-22

Cy-5 direct detection produced higher signals than BTS. Cy-5 direct detection with target amplicon produced a higher signal than Cy-5 non-target control. BTS with target amplicon produced a lower signal than the non-target control.

The leakage in Cy-5 treatments produced very high signals. The BTS system again did not give a change in signal dependent on the presence of target DNA

Cy-5 direct detection produced signals near the same range as the BTS. Cy-5 direct detection with target amplicon produced a higher signal than Cy-5 non-target control. BTS with target amplicon produced a lower signal than the non-target control.

These results show that regardless of BTS use, there is no appreciable signal with the Analyte 2000™ biosensor, showing DNA:DNA interactions are occurring on the waveguide. It was concluded at this stage that there may be nonspecific interactions occurring or the background signal may be too high, resulting in a high signal-to-noise ratio. The next set of experiments attempted to address background issues. Here, the blocking agent was altered, either by increasing the concentration of the blocking agents (Table 2-4); increasing the blocking agent and blocking time (Table 2-5); or increasing blocking agent concentration, time, and incorporating shaking into the protocol (results not shown). Representative results are provided below.

Table 2-4. Effect of Increasing Blocking Agent Concentration on the Signal: Noise Ratio during Detection. Detection of *H. pylori glmM* was performed with experimental blocking agent using 10 pmoles of each probe and 50 pmoles target. The blocking agents were increased from 2% casein, 2% BSA to 4% casein, 4% BSA.

	Non-Target Blank	**Non-Target Blank**	**50 pmoles Target**	**50 pmoles Target**
Initial signal	325	254	437	282
Final signal	291	242	322	277
Δ Signal	-34	-12	-115	-5

Increasing the blocking agent concentration did not increase the target signal over the non-target signal.

Table 2-5. Effect of Increasing the Blocking Time on the Signal: Noise Ratio during Detection. Detection of *H. pylori glmM* with extended blocking time using 10 pmoles each probe and 50 pmoles target amplicon.

	Cy-5 Blank	**Cy-5 50 pmoles Target**	**BTS Blank**	**BTS 50 pmoles Target**
Initial signal	246	7078	294	347
Final signal	239	7183	276	330
Δ Signal	-7	+105	-18	-17

The blocking time was extended from the standard 30 minutes to 60 minutes. The non-target blank signal was reduced and evidence of increased signal with the direct Cy-5 labeled probe was shown. However, the target signal for the BTS system was reduced as well. Incorporating shaking into the blocking phase had no effect, and signals were random irrespective of the presence of target DNA.

Taken together, these results suggest that the high signal: noise ratio was not the result of nonspecific interaction occurring due to the target and/or detection probe interacting with the waveguide. If this was the case, altering the blocking agents and time likely would have altered the signals obtained.

The next set of experiments considered whether the addition of the various DNA components to the waveguide was critical to the generation of the signal. In previous trials, all components were added to the waveguide after denaturation. In the first experiment, the capture probe was added first to bind to the waveguide. This was followed by the addition of the hybridized target and detection probes, followed by the BTS system. Here, there was no appreciable difference in the signal generated with or without target DNA present. This was repeated twice for both target genes. When the capture probe and target were first hybridized together and then added to the waveguide, followed by the addition of the detection probe, again there was no difference in the signal plus or minus target DNA (results not shown). Table 2-6 shows the results when all DNA components were added separately to the waveguide. Here, the capture probe was allowed to bind to the waveguide, the denatured target was then allowed to hybridize to the capture probe, and then the detection probe was allowed to hybridize to the target for 1 hour each.

Table 2-6. Effect of Adding Components for *H. pylori* Gene Target Separately on Overall Change in Signal.

	Blank	Blank	+ Target	+ Target
Δ Signal	+52	+41	+34	+44

Again high blank readings and random signal patterns determined that how the components are added to the system did not affect the signals.

Taken together, all the trials performed on sequence of addition of DNA components to the waveguide did not indicate that this had any effect on the signal generated by the system.

One final experiment analyzing why the signals generated by this system were not DNA-specific concerned the possible nonspecific interaction of the detection probe with the waveguide itself. The FITC-labeled detection probe was allowed to interact with the streptavidin-coated waveguide for 1 hour. Then, the rest of the BTS amplification system (anti-fluorescein HRP antibody, biotin tyramide, and streptavidin conjugated Cy-5) was carried out. Four chambers were run simultaneously with the same treatment. The signals obtained were very low and indicated very little interaction between the detection probe and the waveguide (Table 2-7). The experiments were repeated twice for both detection probes.

Table 2-7. Determination of Nonspecific Binding of the *C. parvum* Detection Probe to the Waveguide.

	Channel 1	Channel 2	Channel 3	Channel 4
Δ Signal	+5	+9	+2	+1

Taken together these results suggest that detection of DNA on the Analyte 2000™ biosensor is not optimal, and that these results have not clearly identified the causes of the problems. However, some possibilities that were not fully addressed and could contribute include:

- Poor binding of the capture probe to the waveguide
- Poor waveguides
- Low sensitivity

The project team addressed the sensitivity by adding more target DNA and did not see a proportional increase in signal. This basically leaves a problem with the waveguides. We started looking at an alternative chemistry for binding the capture probe to the waveguide. Past experiments using a radioactively labeled capture probe suggested that only 1% of the capture probe was binding to the waveguide. This will obviously decrease the interaction of the target DNA considerably and therefore decrease the signal. Thus using a different system to the biotin: streptavidin system might improve this interaction. However, the team was unable to locate a successful alternate method to binding DNA capture probes to the waveguide. In addition, it has become obvious that the problems experienced with the Analyte 2000™ biosensor occurred gradually over the past 6 months or so, raising the possibility that the quality of the waveguides is declining.

To address, the project team looked unsuccessfully into purchasing the waveguides from a vendor other than Research International. The QC of the waveguides is a concern and continues to be an issue. To determine if in fact the waveguides are a problem, we shifted experiments to using a fluorescent ELISA plate reader, which would make use of standard 96 well plates, instead of the waveguides and the Analyte 2000™ biosensor. This set of experiments would allow us to study more parameters at once (rather than the four studied with the Analyte 2000™ biosensor), as well as determine if the BTS system would work. One drawback of the plate reader system is the lower sensitivity.

BTS Performed by Fluorescent Assay on ELISA Plate Reader

The first set of experiments was designed to reproduce the Analyte 2000™ biosensor tests. Detection of both target DNAs was attempted using FITC conjugated streptavidin, 10 pmoles each probe, and target amplicon ranging from 0.5 to 10 pmoles. FITC labeled streptavidin was used here since this is a better fluorescence signal for the ELISA reader compared to Cy-5. Representative results are shown in Table 2-8. These trials were again repeated at least twice for both target DNAs.

Table 2-8. Detection of *H. pylori glmM* Target on the ELISA Reader Using the BTS System.

Target (pmoles)	Capture probe (pmoles)	Detection probe (pmoles)	Reading (rfu)
0.5	10	10	23
0.5	10	10	18
1.0	10	10	17
1.0	10	10	18
5	10	10	20
5	10	10	18
10	10	10	19
10	10	10	20
0	10	10	19
0	10	10	20
0	10	10	14
0	10	10	15
0	10	10	20

No difference was noted between the signal generated regardless of the presence of target DNA. In the next set of experiments the detection and target probe concentrations were changed to see if this would increase the signal when target DNA was present. In all cases, no rise in signal in the target wells over the background (non-target controls) was detected. Using Cy-5 labeled streptavidin also did not improve the signal. Table 2-9 shows that the signals produced were very low and that there was no difference in signal regardless of the presence of DNA.

Table 2-9. Detection of *H. pylori glmM* using Cy-5 Conjugated Streptavidin Standard Protocol with10 pmoles of Each Probe and 0.5 to 5 pmoles of Target Amplicon.

Target (pmoles)	Capture probe (pmoles)	Detection probe (pmoles)	Reading (rfu)
0.5	10	10	0.7
0.5	10	10	1.0
0.5	10	10	0.9
1.0	10	10	1.2
1.0	10	10	0.9
1.0	10	10	1.5
1.0	10	10	1.1
5.0	10	10	1.2
5.0	10	10	1.4
5.0	10	10	0.8
0	10	10	0.6
0	10	10	1.5
0	10	10	0.8
0	10	10	0.7
0	10	10	0.5

Increasing the concentration of the target DNA to 20 pmoles did not increase the signal (not shown). One final experiment to increase the signal involved the incubation of the streptavidin conjugate overnight followed by the addition of the BTS (Table 2-10). Again, presence of target DNA did not make a difference in the signals obtained.

Table 2-10. Detection of *H. pylori glm*M Using Standard Protocol. The final incubation of FITC-conjugated streptavidin was conducted at 4°C overnight with 10 pmoles each probe and 10 and 20 pmoles target amplicon.

Target (pmoles)	Capture probe (pmoles)	Detection probe (pmoles)	Reading after BTS (rfu)	Reading after FITC for 24hr. (rfu)
		Trial 1		
10	10	10	24.3	27.0
10	10	10	26.1	29.0
20	10	10	25.7	25.2
0	10	10	24.1	23.4
0	10	10	18.4	17.2
PBS	0	0	2.2	2.9
PBS	0	0	1.7	1.7
0	0	0	13.1	2.3
		Trial 2		
10	10	10	15.8	15.6
10	10	10	17.8	17.7
0	10	10	20.8	19.4
0	10	10	18.5	17.1
10	10	10	17.1	18.6
10	10	10	20.0	19.8
10	20	20	18.1	18.8
0	10	10	18.2	17.8
0	10	10	20.4	20.7
0	20	20	16.4	20.7

The results using the ELISA plate reader basically mimicked the results obtained with the Analyte 2000™ biosensor. No signal dependent on the presence of target DNA was obtained. At this stage, the project team decided that the BTS system likely was either having no effect or having a deleterious effect on the system. For this reason, subsequent experiments did not use the BTS.

2.9 Direct Fluorescent Labeling Detection System

Direct detection experiments on the Analyte 2000™ biosensor (Note: All analyte assays use Cy-5 labeled probes)

Again, for this set of experiments, at least two trials were performed for each target DNA. Results given are typically representative of the overall results obtained. The first set of experiments used the standard protocol described in Section 1.9 using a Cy-5 labeled detection probe. Table 2-11 shows the results of a typical experiment.

Table 2-11. Detection of *C. parvum hsp70* Using Standard Protocol with 10 pmoles of Amplicon and 10 pmoles Each Probe.

	Blank	Blank	10 pmoles Target	10 pmoles Target
Initial signal	273	357	314	284
Final signal	286	362	316	300
Δ Signal	+13	+5	+2	+16

Very low signals were seen in all channels and there was no signal difference between target and non-target blanks. The same was seen for the *H. pylori* target. Increasing the target DNA concentration to 10 or 20 pmoles did not produce a specific signal. Using 50 pmoles of target DNA and doubling the hybridization time also did not result in a signal (Table 2-12).

Table 2-12. Detection of *H. pylori glmM* Using 10 pmoles Each probe and 50 pmoles Target Amplicon. Comparing a 1-hour (standard protocol) DNA mixture incubation to a 2-hour incubation.

Trial 1	Blank 1 hour	50 pmoles Target	Blank 2 hours	50 pmoles Target
Initial signal	305	334	320	273
Final signal	321	322	334	258
Δ Signal	+16	-12	+14	-15
Trial 2				
Initial signal	286	286	1570	364
Final signal	293	302	1781	409
Δ Signal	+7	+16	+211	+45

As with the BTS system, the project team looked at adding the DNA components separately and together to the waveguide. Again, no difference in the signal was noted regardless of the presence of target DNA (not shown). In addition, no signal was generated at all when target DNA was left out of the hybridization system (results not shown).

Direct Fluorescent Assays on ELISA Plate Reader

It was important to make sure that using the ELISA reader and the FITC-labeled probe would be sensitive enough to detect an interaction, if it occurred. Twelve aliquots of FITC-labeled detection probe used in the BTS system (FITC is used as an antigen for anti-fluorescein HRP antibodies) of 150 μl at 10 pmoles were assayed for bulk signal. The average signal for 12 samples was approximately 2000 rfus. This amount of signal would be acceptable if seen in wells with successfully captured target. However, as was evident with the Analyte 2000™ biosensor, all signals were produced independent of the presence of target DNA. Increasing target DNA, capture probe, or target probe did not change the signal intensity. Results were the same for both target DNAs, and basically agreed with results obtained using the Analyte 2000™ biosensor and the BTS system.

The cumulative results show that this system as it stands cannot be used for the detection of organisms. The project team has exhausted all possible mechanisms to examine what part of the system is not working and why. The next set of experiments continued to use the Analyte 2000™ biosensor, but utilized PCR to increase the signal.

2.10 PCR-Dependent Labeled Amplicon Detection System

These experiments were designed to use PCR to incorporate Cy-5 into the target DNA during an amplification step. This would, therefore, negate the use of a detection probe and thus potentially decrease the interactions required to get a signal. Experiments were performed on both the Analyte 2000™ biosensor and the ELISA plate reader.

Labeled Experiments on the Analyte 2000™ Biosensor

In the first set of experiments, the detection of *H. pylori glmM* was attempted by first subjecting the *H. pylori* genomic DNA to PCR using Cy-5 labeled nucleotides and the primers specific for the *glmM* gene. The synthesized labeled product was then used in the hybridization system on the Analyte 2000™ biosensor. Table 2-13 shows the results of three trials.

Table 2-13. Detection of a Cy-5 labeled Target DNA (*glmM*) Generated by PCR.

Trial 1	**5 pmoles Target**	**5 pmoles Target**	**Blank**	**Blank**
Initial signal	361	352	313	366
Final signal	1584	600	376	356
Δ Signal	+1223	+248	+63	-10
Trial 2				
Initial signal	318	417	403	343
Final signal	336	454	409	515
Δ Signal	+18	+37	+6	+172
Trial 3				
Initial signal	401	330	324	344
Final signal	756	1065	662	883
Δ Signal	+355	+735	+338	+539

All three trials included 5 pmoles of Cy-5 labeled amplicon and 10 pmoles capture probe. In trial 1, higher signals were obtained for the chambers containing labeled amplicon, lower signals for non-capture probe blanks. This run suggested for the first time that a signal was produced that was higher with the target DNA compared to the controls. In Trials 2 and 3, random signals were not representative of the treatments.

These results were initially positive but had poor reproducibility. Repeated trials did not reproduce the same result and consequently were not conclusive. Further experiments added the capture probe first followed by the target DNA did not give results suggesting specific interactions as a result of presence of target DNA.

Labeled Fluorescent Assays Using the ELISA Plate Reader

Initial studies using the ELISA plate reader and a Cy-5 labeled target DNA were somewhat more positive. Using the *C. parvum hsp70* Cy-5 labeled target and the protocol described, results suggested that target DNA presence was indeed increasing the signal (Table 2-14). Here, an increase in signal that was somewhat proportional to the increase in target DNA concentration and resulted from the presence of the capture probe was observed. However, signals were very high, and results are not conclusive at this stage.

Table 2-14. Detection of the *C. parvum hsp70* Internal Cy-5 Labeled Amplicon Using the Standard Protocol with 0.5, 1.0, and 1.5 pmoles of Target Amplicon with 10 and 20 pmoles of Capture Probe.

Target (pmoles)	Capture probe (pmoles)	Reading (rfu)
0.5	10	1018
0.5	20	1082
0.5	0	974
0.5	0	944
1.0	10	1146
1.0	20	1177
1.0	0	979
1.0	0	945
1.5	10	1316
1.5	20	1423
1.5	0	985
1.5	0	952

2.11 Detection of *H. pylori* in Well Water Using Conventional PCR

As an alternative to the Analyte 2000™ biosensor for detection of *H. pylori*, the project team also investigated the use of real time PCR with SYBR green staining as a rapid detection method in order to compare important parameters such as sensitivity, time requirements, costs, and ease of use. Using conventional PCR, the team achieved a level of detection of ≤2 cells/mL. The primers used for these experiments (HP1 and HP2) target the gene for the 16S rRNA. No nested PCR is necessary to achieve this level of detection. Immunomagnetic separation (IMS) was used for well water sampling before PCR because it gives a lower level of detection (LOD), probably due to removal of the cells from PCR inhibitors found in the well water and an increase in cell concentration (Figure 2-11).

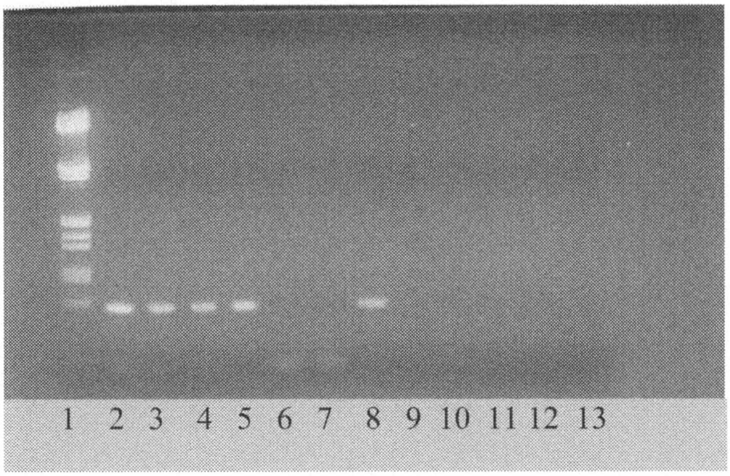

Figure 2-11. Conventional PCR with Primers Hp1 and Hp2. One milliliter dilutions with well water. LOD for dilutions in well water/IMS=1.8 cells; without IMS, LOD=1325 cells.

1. Lambda marker
2. 2.65×10^4 cells/ml, in well water
3. 2.65×10^3 cells/ml, "
4. 2.65×10^2 cells/ml, "
5. 2.65×10^1 cells/ml, "
6. 2.65×10^0 cells/ml, "
7. Negative control (beads + antibody)
8. 2.65×10^5 cells/ml, in well water
9. 2.65×10^4 cells/ml, "
10. 2.65×10^3 cells/ml, "
11. 2.65×10^2 cells/ml, "
12. 2.65×10^1 cells/ml, "
13. 2.65×10^0 cells/ml, "

IMS

In the actual testing of a well, 1 mL samples will not be sufficient to measure bacterial contamination. Therefore, tests were run using 1 L samples of well water spiked with 2.65×10^4 *H. pylori* cells. In order to determine the most efficient way to handle the larger volume, three samples were filtered separately onto 0.2 μm Supor filters and were processed as detailed below:

1. The filter was cut up and placed into two 1.5 mL microcentrifuge tubes with 900 μl of PBS/1% FBS each. The IMS beads were added (20 μl each), and the tubes rotated for 1 hour to allow the beads to target the *H. pylori* cells (Figure 2-12, Lanes 3 and 4).
2. The filter was cut up to fit inside the Dynal tube (15 mL) with 5 mL of PBS/1% FBS added. The IMS beads were added (50 μl). The tube was rotated for 1 hour (Figure 2-12, Lane 5).
3. The filter was placed into 8 mL of PBS/1% FBS, and then sonicated for 30 seconds. The supernatant was transferred to a Dynal tube and 80 μl of IMS beads were added. The tube was rotated for 1 hour (Figure 2-12, Lane 6)

Figure 2-12 shows the results from the 3 experiments above. All experiments included IMS, and the beads were added directly to the PCR mixture. Lanes 3 and 4 represent the contents of one filter, which was separated into two microcentrifuge tubes. The microcentrifuge tubes are much easier to work with and take fewer beads than the Dynal tubes. Lane 5 represents the contents of one filter contained in one 15mL Dynal tube. Lane 6 shows the results of sonication of one filter, followed by transfer of the supernatant to one Dynal tube.

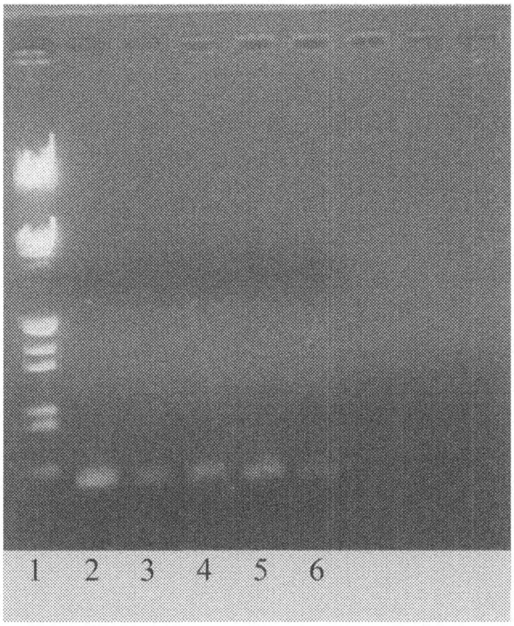

Figure 2-12. Recovery of *H. pylori* Cells from Filter and Detection with Conventional PCR with Primers: Hp1 & Hp2. Lane 1 is the lambda marker and Lane 2 is a positive control. Lanes 3-6 are described in the text above.

The effect of cell preparation method [whole cell, extracted DNA (boiled), or sonicated cells] on amplification by PCR was assessed. One-liter samples of well water were inoculated with ~265 cells. Samples were filtered, and then subjected to IMS followed by the different protocols for template preparation (Figure 2-13).

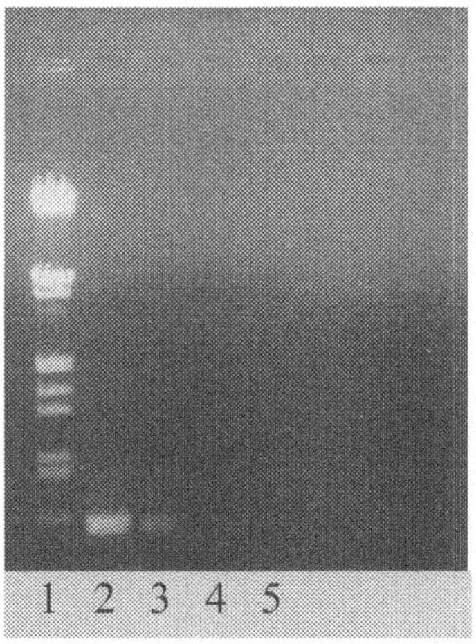

Figure 2-13. Effect of Various Treatments for DNA extraction from *H. pylori* Cells Prior to PCR. Lane 1=lambda marker; Lane 2=positive control (10^7 cells); Lane 3=captured cells boiled; Lane 4=captured cells sonicated; Lane 5=no additional treatment.

The final step was to assess the sensitivity of the assay by determining the LOD following filtration and purification by IMS from 1 L samples of well water (Figure 2-14). The LOD is lower than 265 cells/L, but higher than 26 cells/L.

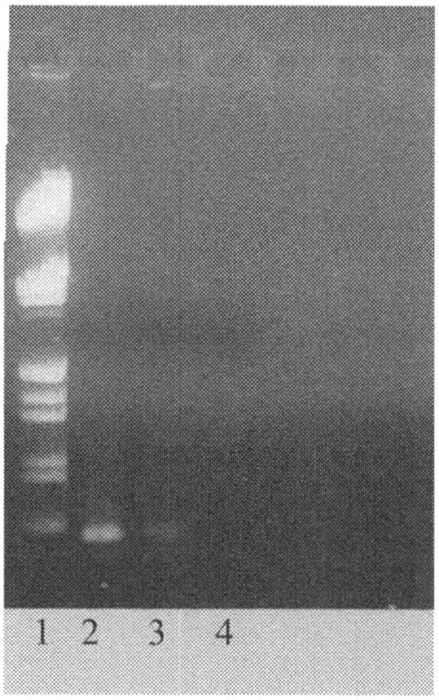

Figure 2-14. Confirmation of Lowest Level of Detection of *H. pylori* in 1 Liter of Well Water. Lane 1= lambda marker; Lane 2= positive control; Lane 3= ~265 cells; Lane 4= ~26 cells.

At this point in the study, the best method for detection by conventional PCR using primers targeting the 16S rRNA is to filter the well water samples and then purify the cells by IMS. The captured cells are boiled to extract the DNA before an aliquot is taken as the PCR template. The limit of detection using this process is at or below 265 cells/L of water.

2.12 Real Time PCR of Well Water Spiked with *H. pylori*

Well water was spiked with *H. pylori* cells to various concentrations. The samples were filtered and *H. pylori* cells were retrieved using IMS, as described in the methods. The final sample was then boiled and used as template for real time PCR with primers targeting the 16S rRNA gene. The results for the real time PCR are shown in Figure 2-15.

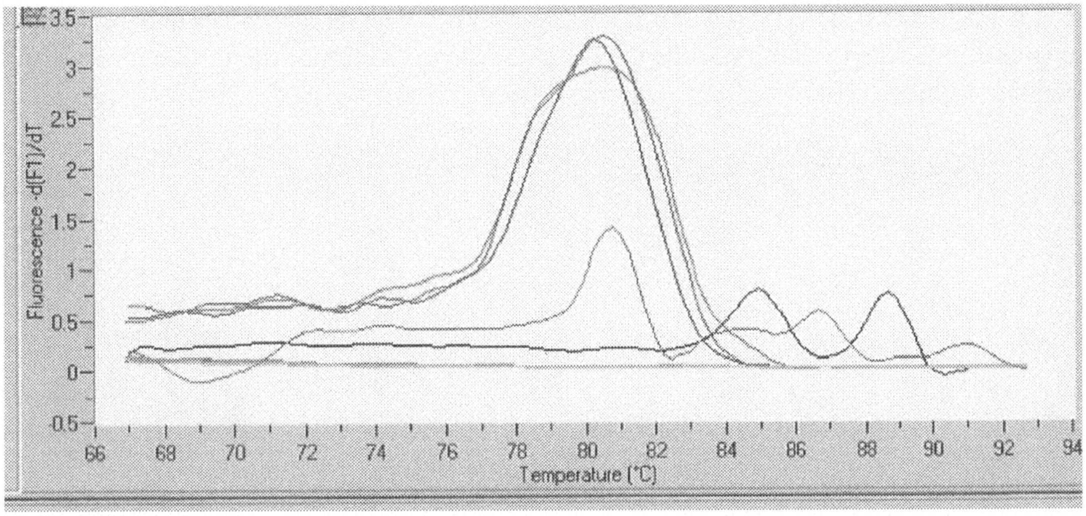

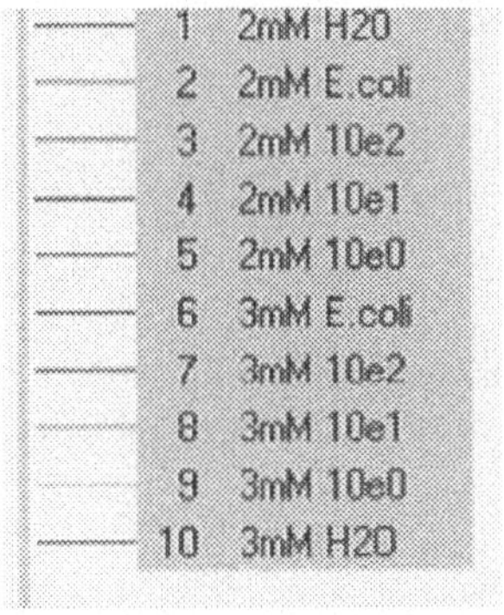

Figure 2-15. Detection of *H. pylori* by Real Time PCR, UsingPrimers to the 16S rRNA Gene. One mL dilutions in well water containing ~265, 26.5, and 2.65 cells of *H. pylori* were exposed to IMS. The captured cells (still attached to magnetic beads) were lysed by boiling and an aliquot was taken for PCR. Sixteen-S rRNA primers were used in replicate reactions containing 2mM $MgCl_2$ or 3 mM $MgCl_2$. The 2mM reaction containing 26.5 cells (2mM 10e1) and one of the negative control produced peaks above 84°C in the melting curve analysis. The specific product has a temperature of 89°C.

While most of the individual reactions gave a signal, melting curve analysis showed that these corresponded to a product that had a differing temperature than the targeted product. These products produced by the negative controls (water or *E. coli* template DNA) had melting temperatures at 84°C, while the authentic product should be 89°C. Only the sample containing 26.5 cells and using 2 mM $MgCl_2$ gave this correct product. This assay was repeated, and similar results were obtained. Since PCR for 265 cells gave the incorrect product, it is possible that the 26.5 cell product was also incorrect. These results suggested that the real-time PCR was not working and that incorrect products/non products were being made. It was decided to use pure DNA and ensure that real-time PCR using these primers would work.

Real Time PCR Using Pure H. pylori DNA

DNA was extracted from a pure culture of *H. pylori* as described above. After quantification, a specific amount of DNA was then used as template in the real time PCR procedure (Figure 2-16).

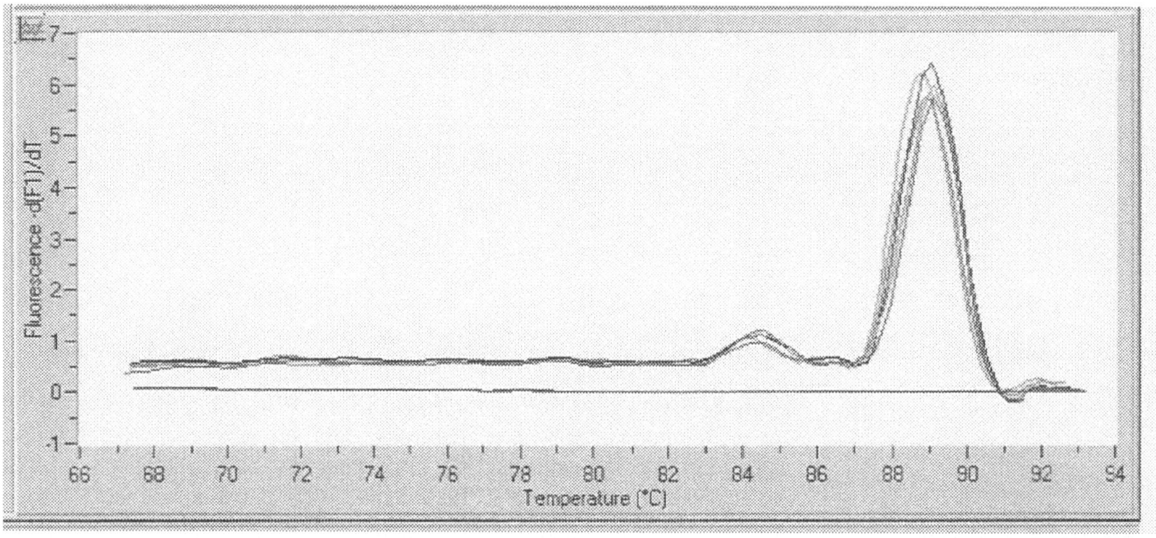

Figure 2-16. Real-Time PCR of Pure *H. pylori* DNA using Primers to 16S rDNA Gene. Dilutions of purified DNA were used for real time PCR analysis. Sixteen-S rDNA primers were used with the following DNA dilutions: 30ng, 3ng, 300pg, 30pg, and 3pg (representing ~10^7 to 10^3 cells, respectively). All show a correct temperature of 89°C.

In this case, the correct product was amplified during the PCR reaction. This product had a temperature of 89°C and was produced only if template DNA was present. Non-template controls gave zero signal. This was repeated using primers to the *ureA* gene (Figure 2-17).

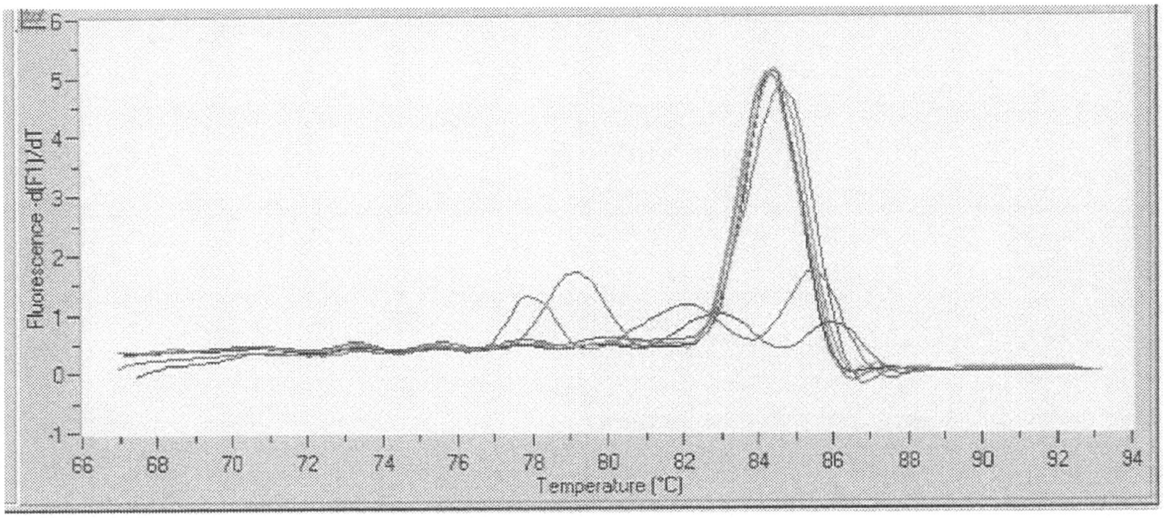

Figure 2-17. Real-Time PCR of Pure *H. pylori* DNA Using Primers to *ureA* Gene. Dilutions of purified DNA were used with *ureA*pPrimers: 3ng, 300pg, 30pg, and 3pg and 0.3pg (representing ~10^6 to 10^2 cells, respectively).

The correct product was made in the samples containing template DNA. This product has a melting temperature of 84.4°C. The two negative controls (sterile water) had various peaks that were not the targeted product at a temperature of 84.4°C. To investigate this further, the products from the real-time PCR were analyzed by gel electrophoresis. The PCR products were extracted from the glass capillaries and then electrophoresed on a 1% agarose gel using 10 µl of product and 5 µl of loading dye per reaction (Figure 2-18).

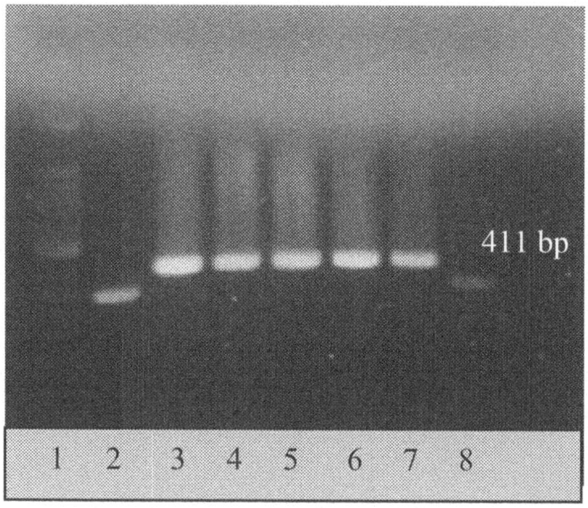

Figure 2-18. Agarose Gel Electrohphoresis of Real-Time PCR Products Made Using *ureA* Primers. Lanes 3-7 show a band with the expected size of 411 base pairs. Lanes 2 and 8 (negative controls) show a smaller product of ~300 base pairs, which could possibly be primer artifacts or a nonspecific product. The sterile water used as template in the negative controls was the same water used in the master mix for all of the reactions. Lane 1=100 bp marker; Lane 2=negative control; Lane 3=3 ng; Lane 4=300 pg; Lane 5=30 pg; Lane 6=3 pg; Lane 7=0.3 pg; Lane 8=negative control

Figure 2-18 shows that the correct sized product (411 base pairs) was made when *H. pylori* DNA was present in the reaction, Lanes 3-7. In the non-template controls (water), a product of 300 base pairs was made, Lanes 2 and 8, which is incorrect and gave the wrong melting temperature. This could be the result of primer interaction or synthesis of a nonspecific product.

Direct Comparison of Real-Time PCR Using Pure Template DNA vs. Whole Cells (16S rRNA Target).

IMS and whole cell lysis was performed on well water spiked with 265 *H. pylori* cells/L, as described in the methods section. The crude DNA mix extracted was then used as template for real time PCR along with pure *H. pylori* DNA. Real time PCR was performed using primers to the 16S rRNA gene (Figure 2-19).

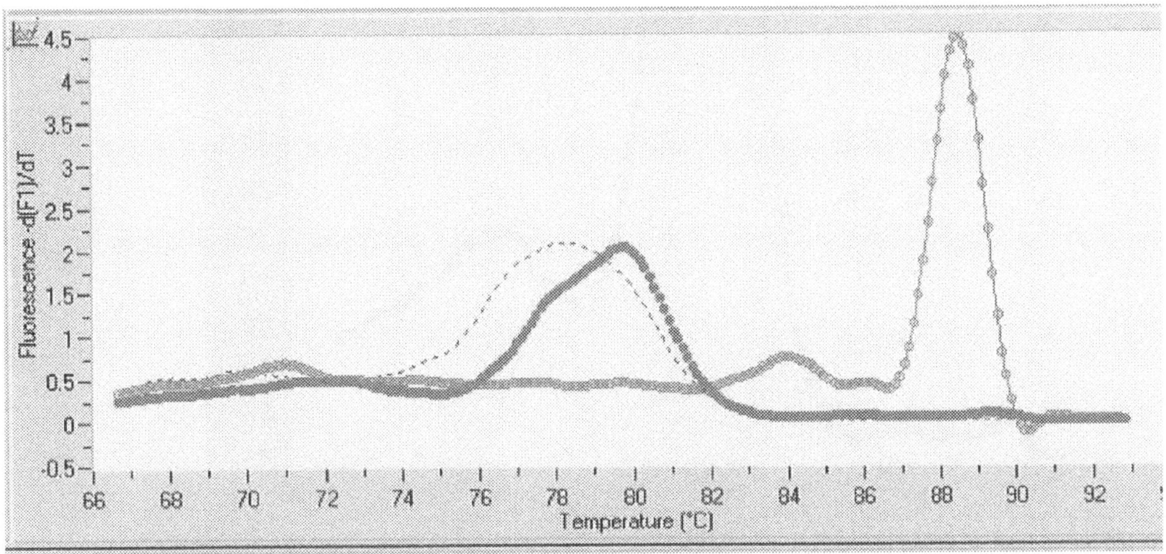

Figure 2-19. Comparison of Real Time PCR Detection Using Whole Cell Lysis and Pure DNA. Primers to the 16S rDNA were used. The purified DNA was 0.05pg ~28 cells. The whole cell assay consisted of 1 L of well water spiked with ~265 cells that was filtered and exposed to IMS. The magnetic beads were boiled to lyse the cells before an aliquot was taken. A temperature of 89°C is shown only for the purified DNA.

Figure 2-19 shows that detection of the 16S rRNA gene by real time PCR was only possible when purified DNA was used as the template. Only when pure DNA was present was the correct product with a melting temperature of 89°C produced.

Interpretation of Results

It has been previously shown that <50 *H. pylori* cells can be detected by conventional PCR using cells that were concentrated by filtration, separated from impurities by IMS, and lysed by a simple boiling step. However, when this approach was used for cell preparation prior to real time PCR, results were negative. Real-time PCR was able to detect <50 cells when purified DNA was used as the template. However, template from <50 whole cells prepared by boiling did not produce a peak with the expected temperature of the PCR product. This suggests that the whole-cell approach, while rapid and sensitive by conventional PCR, will not work with the complex matrix of well water at low cell concentrations. However, detection has been successful with lysed whole cells at $>10^3$ cells/L of well water using real time PCR with SYBR green detection (not shown), and this needs to be repeated.

CHAPTER 3.0

DISCUSSION

3.1 The Analyte 2000™ Biosensor

The aim of this project was to utilize the Analyte 2000™ biosensor as a specific detector of pathogens. DNA:DNA interactions are highly specific; the intent was to use DNA probes as both capture and detection molecules. The antibody: antigen interaction has been successfully used for pathogen detection (Lim, 2001). However, the antibody assay can be limited in some respects, specifically the requirement for antibodies that recognize unique antigens on the pathogen. Use of DNA as the detection tool should have overcome this specificity limitation. Additionally, preparation and assay time would be limited to approximately 2-3 hours. Though this is not real time, it is certainly more rapid than conventional methods currently used for the pathogens studied, such as detection by PCR. By using a biosensor for this detection, it was thought that there would not be a need for DNA amplification by the PCR. While PCR is very sensitive and specific for the detection of organisms, inhibitors and other compounds found in water samples can interfere with the amplification assay and produce interpretation problems. Use of the Analyte 2000™ biosensor for antibody: antigen detection assays (Lim, 2001) has shown that contaminants typically do not interfere with this biosensor assay.

As a direct comparison with the Analyte 2000™ biosensor, the project team incorporated a PCR approach for detection of the pathogens. While PCR has limitations in that contaminants can cause false negative reporting, PCR was be an important side-by-side assessment. In addition, the project team included a real time PCR approach that decreased the assay to 1 hour, as compared to 2.75 hours with conventional PCR.

At the onset the team knew that there was a potential problem with sensitivity of detection using the Analyte 2000™ biosensor. Use of a detection system that targets a genomic sequence limits the number of sites to the number of genomes/cells present. For instance, with 10 cells present for the assay, there would only be 10 targets. This would contrast with a PCR assay where those 10 targets would be amplified, or an antibody: antigen assay, where many target antigens would be present on each of the 10 cells. As such, the project plan was to utilize at least two approaches that would increase the signal. One approach would increase the signal present as a result of positive and specific DNA: DNA interactions. The second approach would amplify the target molecule and signal at the same time. Both directions were compared with a standard DNA: DNA assay that did not involve any signal or target amplification. Consequently, the project required verification that:

a. DNA:DNA interactions could be detected on the Analyte 2000™ biosensor.

b. Such interactions were dependent on target quantity and were target specific.
c. The signal resulting from the DNA:DNA interactions could be amplified such that low target concentrations would be detected.
d. Target amplification could increase both target and signal.

All approaches required extensive and multiple trials to exhaust all possibilities of experimental error. The team identified DNA targets for both organisms; the *glmM* gene (formerly named *ureA*) for *H. pylori* and the *hsp70* gene for *C. parvum*. Rather than use whole genomes and organisms, the team initially used PCR amplified regions of these genes, which assured accuracy in target concentration. In addition, the team was able to directly compare the Analyte 2000™ biosensor with a fluorescence spectrophotometric microplate reader, which essentially measures the same signal as the biosensor, but with less sensitivity. This approach allowed the team to determine where problems likely occurred and provide insight into protocol changes. Figure 3-1 shows a flowchart of the overall scheme of experimentation. The three lines of research are identified: direct detection; BTS amplification of signal; and PCR amplification with signal incorporation. For each direction a standard protocol was determined followed by manipulation of this protocol to adequately identify assay components that adversely affected the signal. It was essential to ensure that any signal being obtained was the result of specific DNA:DNA interaction rather than background signal as a result of poor signal to noise ratio. Discussion of these investigational approaches is best achieved by following each experimental track. In each case, the steps required to achieve detection were evaluated in an ordered fashion to determine where possible problems existed. Consequently, there is overlap of the experimental trials performed. All trials were done using both DNA targets.

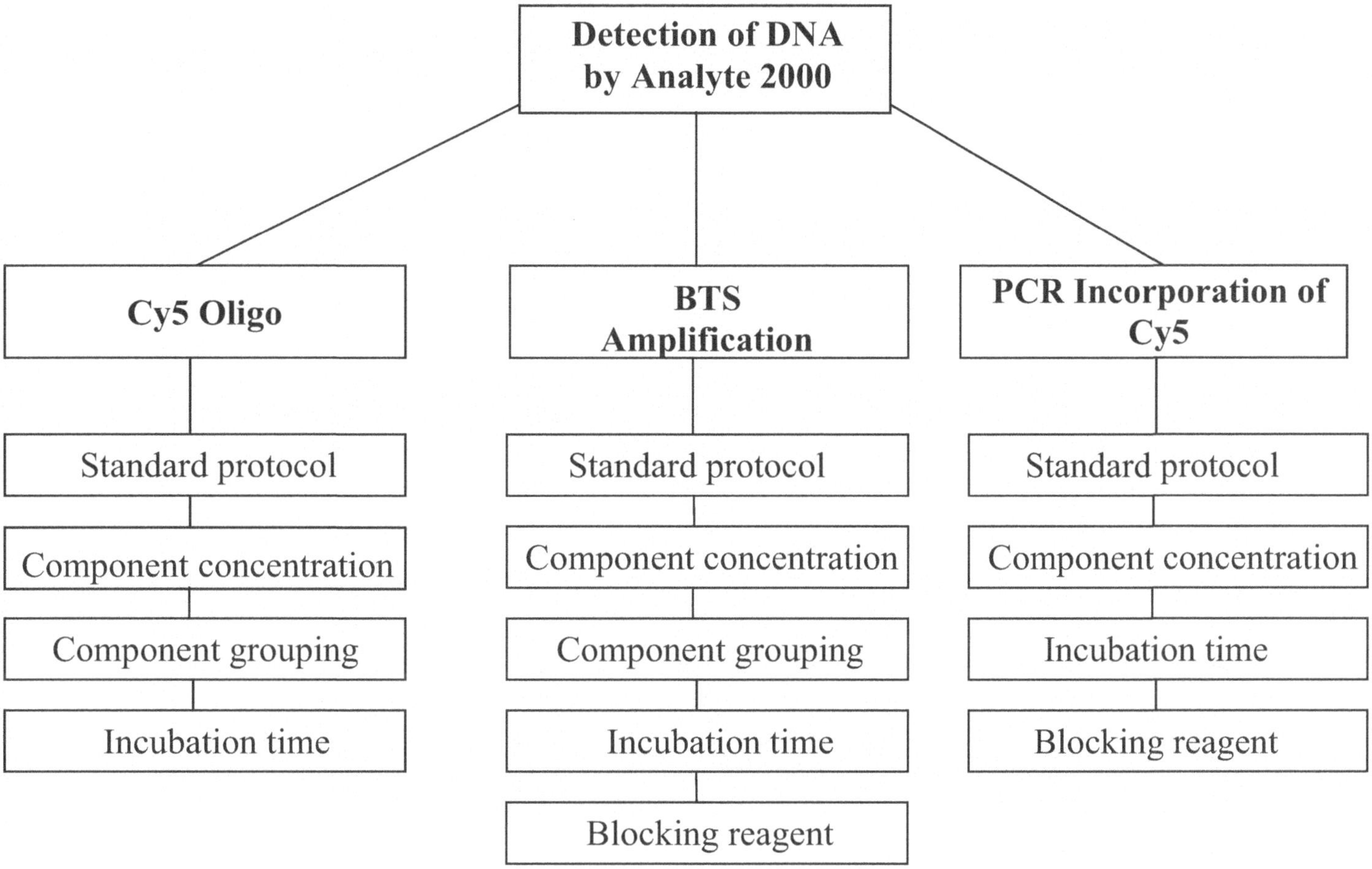

Figure 3-1. Flow Chart Summarizing the Experimental Procedures Used to Evaluate the Analyte 2000™ Biosensor as a DNA Biosensor.

3.2 Direct Detection Approach

In this scenario, the detection probe used included a Cy-5 reporter molecule. This probe should only give a signal if bound to the target molecule (unbound probe is washed through the system), which in turn was bound by a capture probe to the waveguide of the Analyte 2000™ biosensor or the surface of the 96 well microtiter plate. Use of a reporter probe without any amplification limits the detection level. Consequently, it was anticipated that high concentrations of target molecules were required. However, use of the direct detection method avoided any of the additional procedures that were required to amplify the signal and so potentially limited problems incurred by these methods.

Initial studies showed that the target-specific signal was inconsistent. Some trials gave a signal over background and some did not. The inconsistencies suggested that there was a high signal to noise ratio which probably resulted from non specific interactions, either with the waveguide, DNA:DNA, or with components of the assay. These were all considered in attempts to improve the specific signal. Interactions with the waveguide were investigated using isotopically labeled probe, and it was determined that the binding of the capture probe was not optimal. While this does affect sensitivity, it also likely did not contribute to the non-specific interactions. Changing the concentrations, time, and length of addition of the DNA components did not affect the signal obtained. The same experiments were repeated on the fluorescence plate reader with equally poor results. The signals obtained were inconsistent whether target DNA was present or not. The results of the fluorescence plate reader also suggested that though the waveguides had poor capture probe binding, that this might not be the major contributing factor to the inconsistent results obtained.

3.3 Detection Using the BTS Amplification

This procedure took advantage of an enzymatic reaction to increase the number of target sites for binding of a Cy-5 conjugate. This required additional steps compared to the direct detection approach. These steps included the addition of an anti-FITC antibody tagged with HRP, which would bind to the detection probe. Next, a biotin labeled detection probe was applied to the system. Biotin deposits on and in the vicinity of HRP and is thereby responsible for the amplification. Finally, streptavidin: Cy-5 was bound to the biotin. As such, multiple Cy-5 molecules could be specifically bound to the waveguide, increasing the signal.

Use of the BTS was intended to provide a method of signal amplification. While early experiments using this system were encouraging, it ultimately became evident that BTS did not give the desired results. Again, a substantial number of experiments were run to breakdown the various steps of the procedure to determine where the problems existed. These problems were again inconsistent results and very little increase signal as a result of template being present in the assay. Again, the DNA component concentrations, time of addition of components and incubation times were altered in order to improve the signal. In addition, the various blocking parameters were changed, as in concentration, blocking time and incorporating shaking. These protocol changes did not alter the poor signal recorded. With the poor signal that was obtained, new solutions and new DNA components were made, including purchasing new BTS components. Again this did not alter or improve the signal to noise ratios. The comparable experiments done on the fluorescence plate reader were in agreement with the Analyte 2000™ biosensor and showed no increase in signal as a result of target DNA being present.

3.4 Detection Using the Target Amplified with Cy-5

While there was an attempt to avoid using PCR as a procedure for detection, it quickly became obvious that we required an alternative method for signal amplification. Using PCR and Cy-5 labeled nucleotides sought to solve two problems: increasing the number of target molecules and increasing the number of signal molecules.

It was thought that using PCR to amplify a Cy-5 labeled target DNA would solve many of the problems already encountered. This procedure would increase target molecules for capture and would increase the number of detection Cy-5 molecules in the system. It also did not require the steps that were used for the BTS system and thus potentially there was less room for error. Some of the initial trials were promising in that signals from the target DNA samples were higher than the signals from the non-template controls. However, subsequent results were again inconsistent. Use of the fluorescent plate reader method also showed some change in signal as a result of addition of target DNA over non-template controls. The signals produced were extremely high and the increase in signal as a result of presence of target DNA was small and not significant. Again, the effect of the various components and steps were evaluated to determine if the signal could be improved. Altering component concentration, incubation time and changing the blocking reagents and procedure did not affect the overall signals.

At this time the project team cannot recommend using the Analyte 2000™ biosensor as a biosensor for the detection of pathogens at the DNA level. Using very controlled conditions and reagents, we were unable to obtain a detection system that could recognize any level of target DNA. We feel that we fully exhausted all avenues to determine the usefulness of this detection system and conclude that it will not work under these conditions. The aim of the project was to come up with a procedure to rapidly and specifically detect both *C. parvum* and *H. pylori* in water and to add this procedure to complement conventional methods. To this end we have explored the use of the Analyte 2000™ biosensor and have come up short, since this sensor did not meet this aim.

3.5 Detection of Pathogens by PCR and Real Time PCR

To complement the studies with the Analyte 2000™ biosensor, the project team also explored the use of PCR and real-time PCR. Experiences with the biosensor allowed the team to rapidly identify targets and to design and synthesize the required probes for PCR. Additionally, the team had already developed procedures for DNA extraction from the cells. The major initial emphasis was to work with *H. pylori*, and the team developed a procedure using conventional PCR and incorporating immunomagnetic separation that could detect 265 cells/L of water. To speed up this process, the team used real-time PCR instead of conventional PCR. So that the same primers designed for conventional PCR could be used, the SYBR green approach was incorporated, which does not require the specific probe for signal amplification. Initial studies were promising and showed that using purified DNA we could detect 22.5 fg of DNA (approximately 13 *H. pylori* organisms). Using melting curve analysis the products amplified were the correct ones. However, when we used the procedure for template extraction developed for conventional PCR to extract template for real time PCR, the PCR did not function. The products made were of an incorrect melting temperature and gel analysis showed that the products were an incorrect size. This is an important finding since accurate interpretation of real time PCR results is essential for determining presence of target DNA. Obviously, when whole cells in well water are used as template without a DNA purifying step, artifacts confound the accurate detection. Further trials have since helped to improve the detection. Incorporating IMS and increasing the cell number to greater than 1000 cells/L has resulted in the correct amplification of the target amplicon. However, this is at early stages and is being further studied.

3.6 Conclusions

This project set out to evaluate the capacity of the Analyte 2000™ biosensor to detect waterborne pathogens. This biosensor is an evanescent waveguide system that uses changes in fluorescence for detection. Previously, it has been used for detection of chemicals and also cells using antibody: antigen interactions. It seemed a logical step to take this biosensor one step further and utilize the more specific DNA:DNA interactions for detection. This method for detection would rely on capturing target DNA onto the waveguide using specific DNA probes, and then detecting the capture event using a second DNA probe that would initiate the fluorescence signal. Initial studies using this system appeared promising. However, after many trials and experimental approaches this use of the Analyte 2000™ biosensor appears to be inadequate for this type of detection. A major concern that occurred frequently was the high background signal that effectively negated any signal due to specific target DNA interaction. Many changes in the procedure were made to improve this high signal-to-background ratio. This proved to be inconclusive and nothing affected the results. Attempts to amplify the signal after hybridization and attempts to increase the number of target molecules and signals in the target were unsuccessful. The major conclusion is that the Analyte 2000™ biosensor cannot presently be used as a DNA biosensor. (DNA detection is not the intended use of this biosensor, and that the results described here should, in no way, reflect on the biosensor's intended uses.)

A procedure for the detection of *H. pylori* using PCR and real-time PCR was developed. This procedure has the sensitivity to detect low cell numbers (265 cells/L) in well water. There were some issues regarding amplification of the correct DNA region, but these are slowly being resolved. This line of research will continue and will be broadened to include *C. parvum*. This method of detection is not real time since DNA must be extracted. However, it has advantages over the current methods of detection of both these pathogens, in that it is more rapid, easier, less expensive and likely more reliable.

REFERENCES

Bickley, J., R.J. Owen, A.G. Fraser, and R.E. Pounder. 1993. Evaluation of the polymerase chain reaction for detecting the urease C gene of Helicobacter pylori in gastric biopsy samples and dental plaque. *J. Med. Microbiol.* 39(5)**:**338-44.

Chen, X., X.E. Zhang, Y. Chai, W. Hu, Z. Zhang, X.M. Zhang, and A.E.G. Cass. 1998. DNA optical sensor: a rapid method for the detection of DNA hybridization. *Biosensors and Bioelectronics*. 13:451-458.

Clayton, C.L., H. Kleanthous, P.J. Coates, D.D. Morgan, and S. Tabaqchali. 1992. Sensitive detection of Helicobacter pylori by using polymerase chain reaction. *J. Clin. Microbiol.* 30(1)**:**192-200.

DeMarco, D.R., E.W. Saaski, D.A. McCrae, and D.V. Lim. 1999. Rapid detection of *Escherichia coli* O157:H7 in ground beef using a fiber-optic biosensor. *J. Food Protection*. 62:711-716.

Engstrand, L., A.M. Nguyen, D.Y. Graham, and F.A. el-Zaatari. 1992. Reverse transcription and polymerase chain reaction amplification of rRNA for detection of Helicobacter species. *J. Clin. Microbiol.* 30(9)**:**2295-301.

Hirayama, H., J. Tamaoka, and K. Horikoshi. 1996. Improved immobilization of DNA to microwell plates for DNA-DNA hybridization. *Nucleic Acid Res*. 24:4098-4099.

Lim, D.V. 2001. Rapid biosensor detection of foodborne pathogens. *Microbiol. Meth. Newslett*. 18:13-17.

Mascini, M., I. Palchetti, and G. Marrazza. 2000. DNA electrochemical biosensors. Fresenius J. *Anal.Chem.* 369:15-22.

Nikiforov, T.T., and Y. Rogers. 1995. The use of 96-well polystyrene plates for DNA hybridization-based assays: an evaluation of different approaches to oligonucleotide immobilization. *Anal. Biochem*. 227:201-209.

Steemers, F.J., J.A. Ferguson, and D.R. Watt. 2000. Screening unlabeled DNA targets with randomly ordered fiber-optic gene arrays. *Nature Biotech.* 18:91-94.

Stinear, T., A. Matusan, K. Hines, and M. Sandery. 1996. Detection of a single viable Cryptosporidium parvum oocyst in environmental water concentrates by reverse transcription-PCR. *Appl. Environ. Microbiol.* 62(9)**:**3385-90.

Tempelman, L.A., K.D. King, G.P. Anderson, and F.S. Ligler. 1996. Quantitating Staphylococcal enterotoxin B in diverse media using a portable fiber-optic biosensor. *Anal. Biochem*. 233:50-57.

Wadkins, R.M., J.P. Golden, and F.S. Ligler. 1995. Calibration of biosensor response using simultaneous evanescent wave excitation of cyanine-labeled capture antibodies and antigens. *Anal. Biochemi*. 232:73-78.

Wu, Z., I. Nagano, A. Matsuo, S. Uga, I. Kimata, M. Iseki, and Y. Takahashi. 2000. Specific PCR primers for Cryptosporidium parvum with extra high sensitivity. *Mol Cell Probes*. 14:33-9.

Zhou, X.C., L.Q. Huang, and S.F.Y. Li. 2001. Microgravimetric DNA sensor based on quartz crystal microbalance: comparison of oligonucleotide immobilization methods and the application in genetic diagnosis. *Biosensors and Bioelectronics*. 16:85-95.

WERF SUBSCRIBERS

WASTEWATER UTILITY

Alabama
Montgomery Water Works & Sanitary Sewer Board

Alaska
Anchorage Water & Wastewater Utility

Arizona
Gila Resources
Glendale, City of, Utilities Department
Mesa, City of
Peoria, City of
Phoenix Water Services Department
Pima County Wastewater Management

Arkansas
Little Rock Wastewater Utility

California
Calaveras County Water District
Central Contra Costa Sanitary District
Contra Costa Water District
Crestline Sanitation District
Delta Diablo Sanitation District
Dublin San Ramon Services District
East Bay Dischargers Authority
East Bay Municipal Utility District
El Dorado Irrigation District
Fairfield-Suisun Sewer District
Fresno Department of Public Utilities
Irvine Ranch Water District
Las Virgenes Municipal Water District
Livermore, City of
Lodi, City of
Los Angeles, City of
Los Angeles County, Sanitation Districts of
Napa Sanitation District
Orange County Sanitation District
Palo Alto, City of
Riverside, City of
Sacramento Regional County Sanitation District
San Diego Metropolitan Wastewater Department, City of
San Francisco, City & County of
San Jose, City of
Santa Barbara, City of
Santa Rosa, City of
South Bayside System Authority
South Coast Water District
South Orange County Wastewater Authority
Union Sanitary District
West Valley Sanitation District

Colorado
Boulder, City of
Colorado Springs Utilities
Littleton/Englewood Water Pollution Control Plant
Metro Wastewater Reclamation District, Denver

Connecticut
The Mattabasset District
New Haven, City of, WPCA

District of Columbia
District of Columbia Water & Sewer Authority

Florida
Broward, County of
Fort Lauderdale, City of
Gainesville Regional Utilities
JEA
Miami-Dade Water & Sewer Authority
Orange County Utilities Department
Orlando, City of
Reedy Creek Improvement District
Seminole County Environmental Services
St. Petersburg, City of
Stuart Public Utilities
Tallahassee, City of
Tampa, City of
Toho Water Authority
West Palm Beach, City of

Georgia
Atlanta Department of Watershed Management
Augusta, City of
Clayton County Water Authority
Cobb County Water System
Columbus Water Works
Fulton County
Gwinnett County Department of Public Utilities
Macon Water Authority
Savannah, City of

Hawaii
Honolulu, City and County of

Illinois
American Bottoms Wastewater Treatment Plant
Greater Peoria Sanitary District
Kankakee River Metropolitan Agency
Metropolitan Water Reclamation District of Greater Chicago
North Shore Sanitary District
Wheaton Sanitary District

Indiana
Fort Wayne, City of

Iowa
Ames, City of
Cedar Rapids Wastewater Facility
Des Moines Metro Wastewater Reclamation Authority
Iowa City, City of

Kansas
Johnson County Unified Wastewater Districts
Unified Government of Wyandotte County/Kansas City, City of

Kentucky
Louisville & Jefferson County Metropolitan Sewer District

Louisiana
Sewerage & Water Board of New Orleans

Maine
Bangor, City of
Portland Water District

Maryland
Anne Arundel County Bureau of Utility Operations
Howard County Department of Public Works
Washington Suburban Sanitary Commission

Massachusetts
Boston Water & Sewer Commission
Upper Blackstone Water Pollution Abatement District

Michigan
Ann Arbor, City of
Detroit, City of
Holland Board of Public Works
Lansing, City of
Owosso Mid-Shiawassee County WWTP
Saginaw, City of
Wayne County Department of Environment
Wyoming, City of

Minnesota
Rochester, City of
Western Lake Superior Sanitary District

Missouri
Independence, City of
Kansas City Missouri Water Services Department
Little Blue Valley Sewer District

Nebraska
Lincoln Wastewater System

Nevada
Henderson, City of

New Jersey
Passaic Valley Sewerage Commissioners

New York
New York City Department of Environmental Protection
Rockland County Solid Waste Management Authority/ Sewer District

North Carolina
Cary, Town of
Charlotte/Mecklenburg Utilities
Durham, City of
Metropolitan Sewerage District of Buncombe County
Orange Water & Sewer Authority

Ohio
Akron, City of
Butler County Department of Environmental Services
Columbus, City of
Metropolitan Sewer District of Greater Cincinnati
Northeast Ohio Regional Sewer District
Summit, County of

Oklahoma
Tulsa, City of

Oregon
Clean Water Services
Eugene/Springfield Water Pollution Control
Water Environment Services

Pennsylvania
Philadelphia, City of
University Area Joint Authority, State College

South Carolina
Charleston Commissioners of Public Works
Mount Pleasant Waterworks & Sewer Commission
Spartanburg Sanitary Sewer District

Tennessee
Cleveland, City of
Knoxville Utilities Board
Murfreesboro Water & Sewer Department
Nashville Metro Water Services

Texas
Austin, City of
Dallas Water Utilities
Denton, City of
El Paso Water Utilities
Fort Worth, City of
Gulf Coast Waste Disposal Authority

Houston, City of

San Antonio Water System

Trinity River Authority

Utah

Salt Lake City Corporation

Virginia

Alexandria Sanitation Authority

Arlington, County of

Fairfax County Virginia

Hampton Roads Sanitation District

Henrico, County of

Hopewell Regional Wastewater Treatment Facility

Loudoun County Sanitation Authority

Lynchburg Regional WWTP

Prince William County Service Authority

Richmond, City of

Rivanna Water & Sewer Authority

Washington

Edmonds, City of

Everett, City of

King County Department of Natural Resources

Seattle Public Utilities

Sunnyside, Port of

Yakima, City of

Wisconsin

Green Bay Metro Sewerage District

Madison Metropolitan Sewerage District

Milwaukee Metropolitan Sewerage District

Racine, City of

Sheboygan Regional Wastewater Treatment

Wausau Water Works

Australia

South Australian Water Corp.

Sydney Water Corp.

Water Corp. of Western Australia

Canada

Greater Vancouver Regional District

Toronto, City of, Ontario

Winnipeg, City of, Manitoba

Mexico

Servicios de Agua y Drenaje de Monterrey, I.P.D.

New Zealand

Watercare Services Limited

Singapore

Singapore Public Utilities Board

United Kingdom

Yorkshire Water Services Limited

STORMWATER UTILITY

California

Los Angeles, City of, Department of Public Works

Monterey, City of

Santa Rosa, City of

Colorado

Boulder, City of

Georgia

Griffin, City of

Iowa

Cedar Rapids Wastewater Facility

Des Moines Metro Wastewater Reclamation Authority

Kansas

Overland Park, City of

Kentucky

Louisville & Jefferson County Metropolitan Sewer District

Maine

Portland Water District

Minnesota

Western Lake Superior Sanitary District

North Carolina

Cary, Town of

Charlotte, City of, Stormwater Services

Pennsylvania

Philadelphia, City of

Tennessee

Chattanooga Storm Water Management

Washington

Seattle Public Utilities

STATE

Arkansas Department of Environmental Quality

Fresno Metropolitan Flood Control District, Calif.

Urban Drainage & Flood Control District, Colo.

CORPORATE

ADS Environmental Services

The ADVENT Group Inc.

Alan Plummer & Associates

Alden Research Laboratory

Alpine Technology Inc.

Aqua-Aerobic Systems, Inc.

Aquateam–Norwegian Water Technology Centre A/S

BaySaver Inc.

BioVir Laboratories, Inc.

Black & Veatch

Boyle Engineering Corporation

BPR CSO

Brown & Caldwell

Burns & McDonnell

CABE Associates Inc.

The Cadmus Group

Camp Dresser & McKee Inc.

Carollo Engineers Inc.

Carpenter Environmental Associates Inc.

CDS Technologies Inc.

Chemtrac Systems Inc.

CH2M HILL

Damon S. Williams Associates, LLC

David L. Sheridan, P.C.

Earth Tech Inc.

Eco-Matrix

Environmental Engineers International

EMA Inc.

The Eshelman Company Inc.

Finkbeiner, Pettis, & Strout Inc.

Freese & Nichols, Inc.

Frontier Geosciences Inc.

ftn Associates Inc.

Fuss & O'Neill, Inc.

Gannett Fleming Inc.

Golder Associates Inc.

Greeley and Hansen LLC

Hazen & Sawyer, P.C.

HDR Engineering Inc.

HNTB Corporation

HydroQual Inc.

Ingersoll-Rand Energy Systems

Insituform Technologies Inc.

Institute for Environmental Technology & Industry, Korea

Jacobson Helgoth Consultants Inc.

Jason Consultants Inc.

Jordan, Jones, & Goulding Inc.

KCI Technologies Inc.

Kelly & Weaver, P.C.

Kennedy/Jenks Consultants

Komline Sanderson Engineering Corporation

Lawler, Matusky & Skelly Engineers, LLP

Limno-Tech Inc.

Lombardo Associates Inc.

Malcolm Pirnie Inc.

McKim & Creed

MEC Analytical Systems Inc.

Metcalf & Eddy Inc.

MWH

New England Organics

Odor & Corrosion Technology Consultants Inc. (OCTC)

ONDEO Degremont Inc.

Oswald Green, LLC

PA Government Services Inc.

Parametrix Inc.

Parsons

Post, Buckley, Schuh & Jernigan

R&D Engineering/Conestoga Rover & Associates

The RETEC Group

RMC, Inc. (Raines, Melton & Carella)

R.M. Towill Corporation

Ross & Associates Ltd.

Royce Technologies

SAIC Maritime Technical Group

Sear-Brown

Short Elliott Hendrickson, Inc.

Stormwater Management, Inc.

Synagro Technologies, Inc.

Tetra Tech Inc.

Trojan Technologies Inc.

URS Corporation

Veolia Water NATC

Wade-Trim Inc.

Weston Solutions Inc.

Woodard & Curran

WRc/D&B, LLC

WWETCO, LLC

Zoeller Pump Company

INDUSTRY

American Electric Power

ChevronTexaco Energy Research & Technology Company

The Coca-Cola Company

Dow Chemical Company

DuPont Company

Eastman Kodak Company

Merck & Company Inc.

ONDEO Services

Procter & Gamble Company

PSEG Services Corp.

Severn Trent Services Inc.

Shell Global Solutions

Thames Water Plc

United Water Services LLC

Note: List as of 1/31/04

WERF Product Order Form

As a benefit of joining the Water Environment Research Foundation, subscribers are entitled to receive one complimentary copy of all final reports and other products.Additional copies are available at cost (usually $10).To order your complimentary copy of a report, please write "free" in the unit price column.WERF keeps track of all orders.If the charge differs from what is shown here, we will call to confirm the total before processing.

Name Title

Organization

Address

City State Zip Code Country

Phone Fax Email

Stock #	Product	Quantity	Unit Price	Total

Postage & Handling

VA Residents Add 4.5% Sales Tax

Canadian Residents Add 7% GST

TOTAL

Method of Payment:

(All orders must be prepaid.)

❑ Check or Money Order Enclosed

❑ Visa ❑ Mastercard ❑ American Express

Account No. Exp. Date

Signature

Shipping & Handling:

Amount of Order	United States	Canada & Mexico	All Others
Up to but not more than:	Add:	Add:	Add:
$20.00	$5.00*	$8.00	50% of amount
30.00	5.50	8.00	40% of amount
40.00	6.00	8.00	
50.00	6.50	14.00	
60.00	7.00	14.00	
80.00	8.00	14.00	
100.00	10.00	21.00	
150.00	12.50	28.00	
200.00	15.00	35.00	
More than $200.00	Add 20% of order	Add 20% of order	

*minimum amount for all orders

Note: Please make checks payable to the Water Environment Research Foundation.

To Order (Subscribers Only):

Log on to www.werf.org and click on the "Product Catalog."

Phone: (703) 684-2470
Fax: (703) 299-0742.

WERF
Attn: Subscriber Services
635 Slaters Lane
Alexandria, VA 22314-1177

To Order (Non-Subscribers):

Non-subscribers may be able to order WERF publications either through WEF (www.wef.org) or IWAP (www.iwapublishing.com).Visit WERF's website at www.werf.org for details.